Oliver Romberg
Nikolaus Hinrichs

Don't Panic with Mechanics!

Oliver Romberg
Nikolaus Hinrichs

Don't Panic with Mechanics!

Fun and success in the "loser discipline" of engineering studies!

Translated by Joe Steen, Jennifer L. Jenkins
and Carola Kasperek

vieweg

Bibliographic information published by Die Deutsche Bibliothek
Die Deutsche Bibliothek lists this publication in the Deutsche Nationalbibliographie;
detailed bibliographic data is available in the Internet at <http://dnb.ddb.de>.

Dr. Oliver Romberg
oliver.romberg@t-online.de

Dr. Nikolaus Hinrichs
nikolaus.hinrichs@gmx.de

Originally published in the German Language by Friedr. Vieweg & Sohn Verlag,
D-65189 Wiesbaden, Germany,
as „Keine Panik vor Mechanik!, 5. Auflage"
© Friedr. Vieweg & Sohn Verlag | GWV Fachverlage GmbH, Wiesbaden 2006

Translated by Joe Steen, Jennifer L. Jenkins and Carola Kasperek

First edition, April 2006

© Cartoons: Oliver Romberg, Bremen

Editorial office: Ulrike Schmickler-Hirzebruch | Petra Rußkamp

Vieweg is a company in the specialist publishing group
Springer Science+Business Media.
www.vieweg.de

Cover design: Ulrike Weigel, www.CorporateDesignGroup.de
Printing and binding: MercedesDruck, Berlin
Printed on acid-free paper

ISBN 3-8348-0181-X
ISBN 978-3-8348-0181-4 (eBook)

Foreword (no one reads it anyway)

While writing the book, we, the authors, met regularly in Hannover's bar scene for the purpose of discussing mechanics, for brainstorming and especially for other reasons. The most fundamental part was: Every evening a new bar. We were able to derive stimulation for a non-academic presentation of mechanics from these work-related meetings, some of which lasted until the wee hours. We would like to express our thanks to each buyer of this book: We hope, by means of the well-deserved profits, to be able at some point to pay the huge number of remaining open bar tabs.

But yet another book about the basics of Technical Mechanics. *Why? This book has been a long time in coming!* In the forewords of "the others", "simple access to mechanics" has been proclaimed as well as "getting rid of academic rigority" and "giving the reader simple access to the basic ideas of mechanics". What many such authors dream of in their greatest fantasies has become reality in this book! The authors have rigorously eliminated any theoretical blow ups[a]. The fact is often disregarded that mechanics is simply the mathematical description and generalization of everyday observations. On the following pages, the fundamentals of mechanics are presented in a manner that makes them understandable for almost everybody. And again, most importantly: *Reading this book should be fun!* The examples, supported by a lot of cartoons, help to learn by associations and practical experiences. A similar textbook has not existed before - the terms "Technical Mechanics" and "Fun" have always been contradictious. Here, we would like to organize a quick tour through the "building of mechanics" in the most enjoyable manner possible, one that emphasizes the entire beauty and simplicity of the conceptual construction.

Furthermore, in view of the actual worldwide reorganization of lectures in engineering studies (Batchelor & Master instead of various Diploma, Degrees or Graduates), this book contributes well to the international harmonization of learning. The value of other textbooks is, however, in no way to be diminished here. On the contrary: The reading of more advanced academic books is strongly suggested to anyone who would like to get an idea of the solid foundations of the construction and the lovely arrangement of the details.

---

[a] ...which Dr. Hinrichs, regarding the work in question here, often realized with his face contorted in pain.

V

One more thing needs to be made clear here: We did not develop contexts ourselves in this book. We have simply ripped off the content (as far as mechanics is concerned). The sources listed in the bibliography served as a model. The ideas for the exercises are mostly based on the holdings of the Institute for Mechanics at the University of Hannover. But what is new is... well, you'll see! Another external source had to be used for the translation of the book from the original very successful German edition. This work has been done by the native English speaking engineer Joe Steen as well as by Jennifer L. Jenkins and Carola Kasperek...hey Joe and Mrs. Jenkins, Mrs Kasperek, ...thanks for that!

If some of the readers get the feeling that we, the authors, react in a somewhat allergic manner to one another over the course of the book and try as often as possible to get the best of one another, then... really now,... it's most certainly not meant that way![b,c]

And finally, here is a short request: This book is still in an experimental phase. We and the Postal Service always appreciate comments, criticism and suggestions!

Hannover (sometime early in the morning),

Dr. Oliver Romberg
Dr. Nikolaus Hinrichs

---

[b] Dr. Hinrichs would like to expressly emphasize here that despite everything, he finds Dr. Romberg a very sociable, funny short guy.

[c] Dr. Romberg regrets that he is not able to exactly return this flattering compliment, but he would like to emphasize that he considers Dr. Hinrichs a ~~boring~~ remarkable scientist.

Table of Contents

1. In Full Possession of Our Mental Forces And Moments: Statics

Imagine yourself sitting somewhere in a train compartment of the InterRegional Express between the German towns of Aurich and Visselhövede (the latter, located between Hannover and Bremen, can be pronounced much easier after a few shots of tequila, but Dr. Hinrichs can also manage this without the help!).

After spending a few hours in the same position in this cramped seating arrangement, a slight pain in your posterior region will involuntarily manifest itself. This can be attributed to a compressive force. Despite the pain, your willpower to keep from moving within the chosen coordinate system (e.g. rail car No. 234) actually triumphs! And already we have a static system! Statics is namely the science dealing with the action of forces on bodies at rest. And since we didn't have to wait around for Einstein to know that a body under uniform motion can also be considered at rest, the laws of statics can also be applied to bodies or systems moving at constant speed, such as in our example with the train (assuming that it is moving considerably slower than the speed of light).

Figure 1: Above) One Newton pulling, below) Two Newtons pulling

In all areas of mechanics, even the most elementary relationships can sometimes be a real poser because in some way or another they seem to elude our power of imagination. Another reason is the fact that we always try to solve a problem on the basis of our own experience combined with supposed logic. In this respect, mechanics can play some pretty nasty tricks on us (Dr. Hinrichs professes not to have experienced this phenomenon.)

Let's have a look at Fig. 1 (above). It shows a firmly fastened (fixed), massless rope whose free end is being pulled by exactly one Newton (a force of one Newton (1N) corresponds to the force needed to hold up approximately 0.1 kg). Here the rope is assumed to be massless so that we do not have to take any vertical forces into account. The same rope is shown in Fig. 1 (below), but at its other end we now see another Newton (perhaps his equally strong brother) pulling with all his might. Therefore two are pulling. The question now is: In which case must the rope withstand more force? Or in a more mechanical sense: In which of the two cases is the rope subject to a greater tensile force?

According to surveys, those unfamiliar with mechanics are equally divided on this question. One side says that the rope with the two Newtons must withstand more force since, after all, here it is being pulled twice as hard.

The others are convinced that the rope has to absorb the same force in both cases, and that is precisely right! If you cannot grasp that, think about this question: How is the rope in our example supposed to know that in case 1 (below) it is being held at one end by Sir Isaac Newton's brother and not wrapped around a rusty old bollard? The whole thing can also be expressed another way. In case 1 (below) Newton's brother simulates the bollard, having to make quite an effort to keep from being pulled across the lawn by the other Newton. To do so, he must exert the same force as his brother, or as the bollard in the other case. But if this were indeed as obvious as we think it is, then they would not have used four horses to quarter criminals in the Middle Ages but instead would have used just three nags and a tree.

Dr. Romberg, who likes to play hobby philosopher on evenings which turn out to be much too long, asks at this point:

What is a force anyway?

One discovers very quickly that this question is impossible to answer. But it's just great to philosophize about. In science, if something cannot be explained, it is either defined, or dismissed as nonsense for a few decades or centuries. The famous Italian painter, illustrator, sculptor, architect, scientist, technologist and engineer (study hard!) Leonardo da Vinci (1452–1519) has the following definition of force on hand for us [16]:

"I submit that force is an incorporeal faculty, an invisible power occasioned by incidental, external strength, introduced and instilled throughout bodies which appear flattened and shrunken through its use, imparting to them active life of wonderful power; it impels all things of creation to transformations in their shape and situation, rushes impetuously to its desired death, altering itself according to cause; made great by delay, weak by swiftness, it is born of strength, and freedom is its demise ."

You really could add the unregistered trademark "poet" to the above list. For the mere mechanical engineer, a force is by definition *a phenomenon which either causes motion or deformation, or restrains it.*

3

Let's imagine, for instance, someone pressing their nose quite emphatically against the side panel of a heavy, free-standing cabinet. At some point in time the test person will begin to detect deformation, or even motion, namely at the moment when the cabinet with its blood-splattered side panel suddenly tips over. This is all due to a force being applied (as mechanical engineers say) by the nose to the cabinet and simultaneously by the cabinet to the nose. When determining the forces at work here, it does not matter which side of the problem we deal with: The forces on the side panel and nose are basically equal in magnitude but act in exactly opposite directions. Voilà! We have already grasped the first and most important "axiom" of statics:

Action = reaction (force = counterforce).

In mechanics this axiom is also known as "Newton 3": "The forces of action and reaction are always equal, or the resultant forces of two interacting bodies are always equal in magnitude and opposite in direction."

(The only exception to this is when Dr. Hinrichs winks at some dolled-up babe...).

If the rope example (Figure 1) is treated professionally as a simple exercise in statics, it first has to be drawn in a simplified manner. We simply symbolize the tugging figure by an arrow. Because a force is actually a vector, i.e. you have to specify not only its amount (here: $F = 1$ N) but also the direction in which it acts. The action of a force is thus a function of its magnitude and direction.

Figure 2: Equivalent system for "Newton pulling a rope"

Instead of the stationary bollard, we now use an equivalent mechanical model: the fixed support. Figure 2 thus shows a complete equivalent mechanical model for the system: "One Newton pulling on a fastened rope."

1.1 The Most Important Thing of All...

The most important thing in the life of mechanics students of both sexes starts with an "*F*" and ends with a "*g*". Something you simply can't get out of your mind! Unfortunately, it is practiced far too little and only in very few cases is it performed with the necessary care and love, thus leading to all sorts of problems. But it does not have to be that way if you learn how to do it properly from the very start: free-body drawing, that is.

5

Drawing a free-body diagram means isolating a system (here the rope) from all figures, bollards, levers, weights, etc. and replacing them with arrows (forces and moments). Here you want to make sure that the system no longer has any contact with its surroundings when the free-body diagram is finished.

So – much like using a pair of scissors – we cut the rope free of all connections, isolating it from its surroundings (bollard and person). And wherever we cut through material (here the rope) we have to draw in (at least) one force. We are allowed to cut wherever we want! Thus we can determine the forces at any point of a body (even within it), provided that they have been correctly indicated. Here the choice of direction of the forces is not crucial. It is much more important to consider whether the point at which we have cut is acted upon by a horizontal and/or vertical force and/or a moment (discussed later). If anything is overlooked in this regard, then the determination of forces and moments goes down the drain. Thus, success in solving a problem or exercise in mechanics rises and falls with taking a close and precise look at it.

So if we remove the Newton figure, we must then replace it "right away" with an arrow (force) to keep the rope from going slack. This has already been done in Figure 2. But the same also holds true for the bollard or for the other Newton. When drawing a free-body diagram it makes no difference whether the forces are applied externally (person) or "originate" in the system (bollard, or even better: support). When dealing with externally applied forces you should make sure to note carefully the direction of the arrow from the start. For a rope the direction is simple, since a rope can transmit tensile forces only. This is clear to anyone walking their dog who has ever tried to push it with the leash. For the forces originating in the system, which are the ones we are usually interested in, the sense of arrow direction does not really matter for the moment. The mathematics of the calculation will let us know if we have drawn it correctly!

So isolating the two cases in Figure 1 results in the same free-body diagram (see Figure 3) because both systems are mechanically identical:

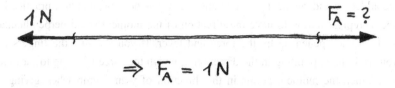

Figure 3: Free-body diagram of "Newton pulling a rope"

In statics, 90% of the more complicated systems differ from one another only in that they comprise more forces (generated by figures, weights and other force providers), moments (coming up next) or supports, and that they have to be treated in terms of two or three dimensions. So for each direction, all arrows must cancel each other out as a whole, since nothing is supposed to be moving (statics). Forces and moments are just summed up completely independent of one another. That's all.

The paradoxical thing about the free-body method is that it is simply not taken up and applied by beginning students of mechanics. The reason for this is one of the last great mysteries in elementary mechanics.

1.2 Just a Moment !

A body can be acted upon by not only forces but also by moments, which tend to cause a turning (rotational) motion instead of a linear (Dr. Hinrichs prefers to call it "translational") one. The best way to imagine a moment is to think of it as a force pulling or pushing on a lever. One and the same force can result in completely different moments, depending on the length of the lever (lever arm). The unit of moment is 1 Nm, due to the fact that the magnitude of the force is multiplied by the lever arm. When determining the moment on the fly without using vector calculation (coming up later), remember that force and lever must be perpendicular to one another.

In statics, the process of determining the moment almost always involves nothing more than applying the "law of levers", which is also well-known to Mr. and Mrs. Moms everywhere, except that sometimes more than just one lever and two forces are involved. Important: Forces and moments in a system are added up independently of one another. It has become standard practice in vector representation to have the direction of the moment stand perpendicular to the plane opened up by the force and lever. If you imagine the fingers of your right hand pointing in the direction in which the force is trying to turn the lever, then the moment points in the direction of your thumb when giving a Michael Schumacher-like "thumbs up" sign[1] ("right-hand rule").

Dr. Hinrichs would like to point out that the drawing of the moment in Figure 4 is not quite right, for when a moment M is replaced by a force F and a lever L, all of a sudden you have a resultant force. That is not supposed to happen in statics, otherwise something will start moving.

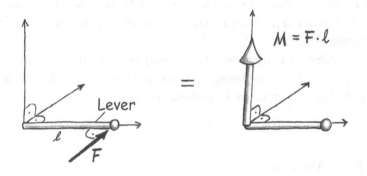

Figure 4: Moment

Here mechanical engineers would say that the equation for the equilibrium of forces ($\Sigma F = 0$, still to come) has not been satisfied (Dr. Hinrichs, however, finds his greatest satisfaction in equations, which is something you will just have to accept. He was a little too much into science friction). A moment is therefore, if it occurs at all in a test problem, replaced by a couple, see Fig. 5.

---

[1] Note: Giving the Rockefeller "finger" is not much help here!

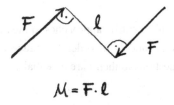

$$M = F \cdot \ell$$

Figure 5: Couple of forces = moment

Here both forces cancel each other out, since a moment by itself does not cause any translational motion. The effect of the moments, however, remains the same. This relationship will become clearer when you start working through the first simple examples below.

For the most beautiful moments in live!

1.3 First, Find a Common "Denominator"

Before we really launch into statics, let's first explain a few fundamental concepts and make a couple of simplifications so that we can conduct an "intelligent" discussion of the problems involved.

1.3.1 The Rigid Body

As is the case in almost all scientific approaches to the (apparently)[2] real world around us, one resorts to models in order to reduce the formulation of the problems and related questions to their bare essentials.

No system, no matter how simple, can be grasped in its entirety by scientific methods. For to do that you would not only have to know the coordinates of all elementary particles at any given time, but also all temperatures, emissions, specific motions, transformations, etc. If you really wanted to establish such knowledge definitively, you would be better off

---

[2] The term in parentheses were added at the express behest of Dr. Romberg.
[3] At this point the editor would like to make the important note that the sun as drawn in its low position has an oval deformation....

joining some harmless religious community! But if you want to describe nothing more than just the forces and moments acting on a system, then you can leave out almost everything else. A model is perfectly sufficient for approximating reality.

One such model is the *rigid body*. Being infinitely stiff and firm, it cannot be deformed even under the greatest of forces. In the technical field, this is usually an admissible and extremely sensible simplification for working with any stable structural component. In reality members do deform when subjected to forces, and that means that the points where the forces are applied shift as well, which in turn means that the relationship of forces acting upon a body can change. This unfortunate condition is completely disregarded for a rigid body. We will therefore assume that the geometry of the system remains constant no matter which forces are acting on it.

1.3.2 The Geometry of Forces

The site upon which a force acts is called the *point of (force) application*. This point, in conjunction with the direction of force, defines a line which is referred to as the *line of action*. In statics, it is permissible to shift a force back and forth along this line of action without altering the system, as long as the direction and magnitude of force are maintained.

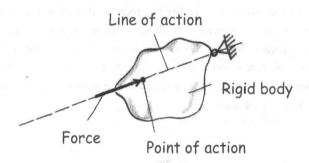

If you wish, you can test this with a simple experiment: Place any rigid body, such as a vase or simply a drinking glass, on a shelf against the wall (support). Now press against the middle of the body with your finger, making sure it does not slip to either side, thus ensuring that it remains at rest due to the acting forces of the finger and the wall (!). Mechanical engineers would say that the body is in a state of *equilibrium*.

The object will still be at equilibrium, with the resultant action and even the forces remaining the same, if we decide to shift the external finger force to another position along its line of action. To do so, we simply take a shaft of any length, for example, an examiner's car antenna which has accidentally broken off a pencil, holding its length between our finger and the object. By maintaining the same magnitude and direction of forces, the effect (equilibrium) on the original object remains the same.

In statics, equilibrium means that a body does not move as a result of the forces and moments which may be pushing or tugging at it from all directions. The forces and moments just cancel each other out in all directions. In the case of a rigid body acted on by only two forces, the two forces must lie on the same line of action, be equal in magnitude and opposite in direction. These conditions are precisely satisfied in our example of the glass against the wall.

The three forces of a central force system can always be summed up by what is referred to as a *force polygon* (which for you science fiction freaks does *not* refer to a character from *Star Wars*!).

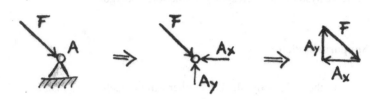

Figure 6: Force polygon

If three forces act on a rigid body, they must – besides obeying the law of "mutual cancellation in every direction" – intersect at a single point. Exception: They are parallel (in which case they make contact with each other in infinity[4]). If a number of forces pull or press upon a body, the outcome can be represented either graphically (e.g. parallelogram of forces) or geometrically (vectors) by a *resultant* (see Fig. 7). In statics the resultant must comprise all forces emanating from the "zero vector".

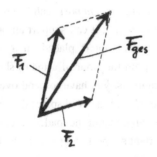

Figure 7: Resultant force

---

[4] Dr. Hinrichs thinks this is very romantic...

A resultant force has the same external effect on a rigid body as all of the combined forces it comprises. Thus it is possible to replace a bunch of forces by their resultant.

A *Cremona diagram* (used mostly for trusses, which we will discuss later) refers to a closed chain of force vectors obtained from an arbitrary number of external forces acting on a system in equilibrium. For three forces of a central force system this corresponds to the force polygon.

1.3.3 Supports in Plain

In addition to the rigid body, there are other little models in mechanics which make life a lot easier for the experienced math whiz. The structural components we have been considering, which keep a loaded object in equilibrium and consequently exert reaction forces of their own, can be regarded as various kinds of supports. The easiest kind of support to imagine is the *pinned support* (which is *not* a new wonder bra for the chastity-minded!). So, for example, if you nail your mouse (computer mouse, this is just an imaginary model!) to your desk, it can no longer execute any translational movements, but just rotate. When lateral pressure is applied, the nail can then transmit two force components in the x and y directions. It is a two-dimensional support.

The one-dimensional *roller support* can compensate forces in one direction only. Like a smooth wall, it can support an object but cannot keep it from sliding to either side. An object placed on a roller support can slide around like a piece of soap on the floor of a men's shower room where, after taking special safety precautions, you have to bend over to pick it up. A roller support can also be thought of as a displaceable pinned support which, depending on its location, can transmit the loaded forces in one direction.

The third most important support in a plane is the *built-in support*, a fixed support secured to keep from rotating. This arrangement can be regarded as a rod whose one end is tightly secured on the edge of a table with a C-clamp.

The rod can neither be displaced in a translational manner nor rotate around the support. It can therefore not only transmit forces in the x and y directions but also compensate for any moments which might occur.

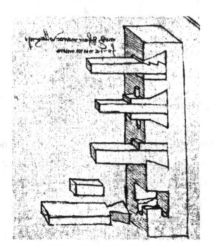

Study by Leonardo da Vinci for
a (displaceable) fixed support [16]

It therefore has a three-force value. This type of support can also be rendered as a so-called sliding sleeve, as a roller support restrained from rotating, or even as a moment support, which transmits only moments which arise (hard to imagine and hardly ever occurs).

The following table provides a quick rundown of the force values and symbols of the most important supports introduced above:

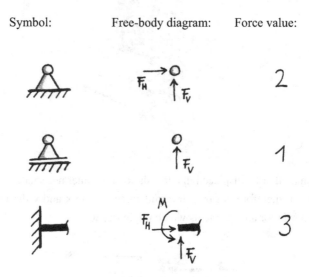

Symbol:	Free-body diagram:	Force value:
	$F_H \rightarrow$ $\uparrow F_V$	2
	$\uparrow F_V$	1
	M $F_H \rightarrow$ $\uparrow F_V$	3

Table 1: Commonly used supports

1.3.4 Other Helpful Models

Rod: A rod is the model idea of a thin, rigid bar with a pinned joint located at each of its ends. This kind of pinned joint is practically like a pinned or roller support, but which never has to be attached anywhere. The end of another rod can just as well be located at this often free-floating joint.

A number of rods which are always connected to each other by their pinned joints will give you a truss (we'll get to that later). If forces only act on the joints, the rod can transmit only one force in the direction of its longitudinal axis, with the direction of force at the supports already being determined (verrrrry important!). In contrast to a beam (see below) a rod is a structure which can transmit only tensile and compressive forces. You can also think of it as a frozen rope. A rod has a completely one-track personality, something which can distribute only tension or pressure. (Taken originally from the word "staff", it is most likely a derivation of "staff officer".)

Pendulum support: A rod that has been built-in somewhere and is subjected to forces only at its ends is called a pendulum support.

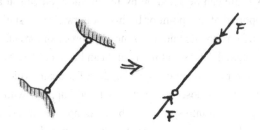

Fig. 8: Rod

A pendulum support can basically only absorb forces in the direction of its longitudinal axis, which means that the direction of force at its ends (joints) is known.

So if a pendulum support is built-in somewhere, all you have to do to determine the direction of force at the isolated rod (and at the intersecting face of the counterforce (action = reaction)) is simply draw a straight line between the end points of the (perhaps even crooked) rod.

Beams: A beam can be acted on by forces and moments at any point and can also be supported at any point or built-in somewhere. In statics, this model is also rigid, which means it is not deformed by forces and moments. However, internal stress (internal forces and moments) can be calculated by isolating it in an appropriate free-body diagram. One model of a beam which is frequently used is the cantilever, whose one end is a fixed support, with its other end projecting obscenely out into space. The best example of an (elastic) cantilever is the diving board at the swimming pool with its common shower room.

Pulley: In statics, a frictionless pulley merely serves the purpose of getting some forces around a corner. For example, with a pulley you can deflect the weight force into the horizontal or in another direction (see the following figure).

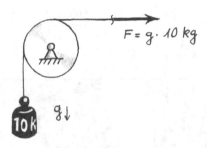

In the following, we will be dealing exclusively with rigid bodies which, due to the forces and moments acting on them, are in a state of equilibrium. For simplicity's sake, we'll first consider systems in a single plane, or in two dimensions (3D functions analogously).

A force vector thus has only two components (direction segments), in the x and y directions (plane of the page), which can be considered independently of one another. In the case of a plane there is, in addition to the two aforementioned translational directions x and y, the possibility of rotation in the direction of angle φ. A moment that rotates around φ "points" in the z-direction (out of the plane of the page, the third coordinate axis in the Cartesian 3D system).

So, did we lose you?... Dr. Hinrichs is foaming at the mouth again... but now we'll slow things down a little.

1.4 Thou Shalt Determine Support Reactions

In almost all cases, the first thing you have to do is calculate the exterior support forces and support moments, in short the "support reactions" of a system involving externally applied forces and moments, which are usually known. In our first example (see Fig. 9) we see a welder's apprentice struggling with a rusty bolt at the free end of a massless cantilever having length L. Fortunately, he is being provided with expert support by his foreman, a certified engineer (in mechanical engineering).

Fig. 9

Here we have made the bold assumption that the steel girder is massless. One thing you should know is that in the technical "sciences" you are allowed to make all sorts of arbitrary assumptions as long as you have a good reason for doing so. Our reason here is that we do not (yet) know where a possible weight force of the girder might act.

To determine the support reactions in the next step, we must first isolate the system and draw in all forces and moments correctly, or in other words, we must first draw an accurate free-body diagram. For determining the support reactions we must first and foremost draw a free-body diagram. The most important thing at the very start is to draw a correct free-body diagram.

Before we even start to consider the problem further, we first draw a free-body diagram. We first draw an intelligent free-body diagram before we start calculating the support reactions. At the start of every statics problem you first draw a free-body diagram. The first thing we do is to isolate the system! We draw a free-body diagram. Even before we have really understood the problem in the first place! The first thing we do is draw a free-body diagram. At the start of every statics problem we therefore draw a correct free-body diagram. First we isolate the system by cutting it free. Isolate – the most, and we mean the very most, important thing of all...

So in order to make the aforementioned free-body diagram, we replace our welder with a simple weight force F, which is known to act in a downward direction and tends to take everything else down with it:

As already mentioned above, we have simplified the system by making the girder massless. Now let's imagine a pair of scissors, or even better, a flex, and, to the energetic protest of the two specialists shown, cut the girder free, isolating it from its support. The girder, now suspended in mid-air, is ready to crash any moment due to the weight of the welder (the girder itself does not weigh anything, of course). However, we can instantly calm the specialists' protest by replacing the support with equivalent support forces. In reference to Table 1, we take the equivalent support force reactions, three in number, and draw them at the point where we cut out the system. Now the system is the same as it was before, at least in terms of statics.

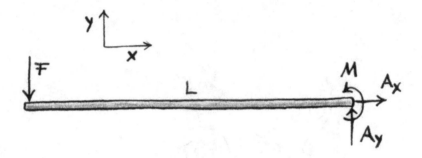

Fig. 10: Free-body diagram for the externally loaded girder

But when drawing a free-body diagram, you must be extremely careful not to forget a force. If you use a Cartesian, or rectangular, coordinate system, the x axis usually coincides with the horizontal forces. As pointed out a number of times already, the conditions of equilibrium demand that the summation of forces and moments must cancel out for every direction.

We first consider the x direction and ask ourselves the following question: How large does the *horizontal* (index H) support reaction $F_H = A_x$ have to be in order to cancel the summation of forces in the x direction? Answer: "zero!" (you dope...). For the *vertical* forces (index V), one comes to the conclusion that $F_V = A_y$ must have the exact magnitude as the weight force F of the welder. The fixed end moment M, in terms of its magnitude, can only be F times L. Only a few forces are involved in this system, and since they also correspond to the coordinate directions, it is pretty easy to determine the reactions without having to write down the individual summations. But to avoid making casual mistakes, it's a good idea if you start getting into the habit of formulating for any system, no matter how simple, the equations for the force summations (Σ) of each direction according to the following method (but not until you have drawn a free-body diagram, first comes the free-body diagram! (see Fig. 10)).

In terms of the coordinate system, all forces which point in the coordinate direction are considered positive (+), while all forces which point in the opposite direction are given a minus sign (−).

Thus, the force summation for the x direction (horizontal) is:

$$\Sigma F_x = A_x = 0 = F_H \implies F_H = 0 \ ;$$

and for the y direction (vertical):

$$\Sigma F_y = 0 = -F + A_y \implies A_y = F = F_V.$$

For determining the momentum we have to establish not only the "rotational direction" but also decide on a point of reference around which the "rotation" takes place. Here the point may lie somewhere in the universe (i.e. in the two-dimensional case somewhere in the infinity of the planes...). But it always makes sense to put the point of reference in the system under investigation. Here's a good trick you can use:

Choose the point of reference so that as many unknown forces as possible do not have a lever arm (usually a support)!

Reason: The moments of forces having no lever arm are zero and therefore do not have to be included when summing up the moments. So, these unknown forces don't appear in the equation of moment summation. All moments pointing in the z direction, or out of the plane of the page ("right-hand rule"), are counted as being positive. We thus form the summation of moments around the support point A, because this is where the lever arm for the most unknown forces disappears:

$$\Sigma M_z^{(A)} = 0 = F L + M \implies M = -F L.$$

The minus sign in front of the term F L tells us that in our free-body diagram we drew the fixed end moment M the "wrong way around", which of course we didn't know at the time. That doesn't matter! It's no mistake! The mathematical expression will tell us which way it has to go. Having the solution, we now know that the fixed end moment has to turn the other way around in order to keep the girder with the welder on top on an even keel. The direction of the support reactions in the free-body diagram is neither here nor there.

Taking a different point of reference, we also arrive at the same result. For instance, let's take the midpoint of the beam:

$$\Sigma \, M_z^{(midpoint)} = 0 = F \; 0.5 \, L \; + \; A_y \; 0.5 \, L \; + M \, .$$

The lever arm is already included in the fixed end moment M, i.e. we do not necessarily have to draw in this external moment directly at the point of restraint, although that is where it provides us with the most useful overview. We could equally well assume the moment M to be in the vicinity of never-never-land, provided that it (in our case) is located on the same plane as the beam. An external moment may be shifted arbitrarily in the plane, a curiosity which for beginning students really takes a while to get used to, like so much in mechanics. This possibility of shifting is best illustrated by the fact that the moment in the "rotational equation" is just written as the letter M, with no reference made to its location. The forces, however, are locally dependent on the lever arm, which itself is dependent on the center of rotation (reference point). And it really does work this way: If you start to drill through a board that's tightly fastened to a workbench with a screw clamp or even nailed to it, the moment exerted on the screw clamp, or the force transmitted to the nails, is completely independent of the actual location of the drill bit (external moment)!

Let's go back to our beam and its new reference point:
Since $F_V = A_y = F$ (see above), it follows that:

$$F \; L \; + M = 0 \; \Rightarrow \; M = - F \, L.$$

As an advanced student of mechanics, you often run short of an equation and things don't sum up. But don't be fooled. Selecting a new point of reference does not provide you with a new equation! On the other hand, you can also determine such a system by using two moment equations and one force equation. But in that case the two points of reference may not be located on a line lying in the direction of the substituted force equilibrium. It's even possible to use three moment equations. In this case the points of reference may not be collinear at all. Of course, this is something you can try out if you are in the mood to do so, or as in the case of Dr. Hinrichs, this puts you in the mood in the first place (!).

The example we have just dealt with is pretty straightforward, since the few forces involved all point in the direction of the coordinates. But normally, due to given natural situations or the sadistic designers of mechanics problems, this cannot be assumed. So now let's look at an example where forces act in "skewed" directions, meaning that they have to be resolved in their coordinate directions. Figure 11 shows a frying pan (mass m, frying radius R) hanging on point A by its massless handle (length L=2R) and held in place at point B by a rope S. The pan is also acted upon by a large, known, and thus given, force F in the manner shown.

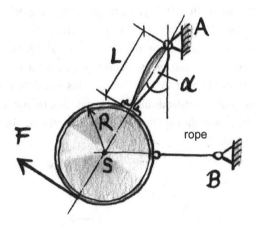

Fig. 11: Hanging frying pan

This example has been specially selected so that you can get used to the deeper meaning of these kinds of exercise problems as early as possible. We want to figure out the force acting on the point of suspension A. So here we are back to calculating support reactions. And to find these, we must first isolate the system, correctly drawing in all forces and moments, or in other words, we have to first draw the right free-body diagram. In order to determine the support reactions, the very first thing we do is draw the proper free-body diagram. Before we even start to consider the problem further, we first draw a

25

free-body diagram. We first draw an "intelligent" free-body diagram before we start calculating the support reactions. At the start of every statics problem you first draw a free-body diagram.

The first thing we do is isolate the system! We draw a free-body diagram. Even before we have read through the problem a second time! The first thing we do is draw a free-body diagram. At the start of every statics problem we therefore draw a correct free-body diagram. First we isolate the system by cutting it free. Isolate – the most, and we mean the very most, important thing of all.

When isolating a free-body diagram for determining external reaction forces, it makes no difference at all what the system looks like internally (reeeaaallly important). There may be ropes, springs, hinges and all sorts of soft tissues built into it. In effect, you can just throw a black cloth over the system and then start isolating it at your leisure. This is allowed when determining external forces, *since the system in equilibrium has assumed precisely this form as a result of the action exerted on it by precisely these forces.* (<= It's probably best to read through this sentence one more time). Anyone who does not understand this can always take refuge in more elegant scientific wording, postulating that "this demonstrates the principle of rigidity... ahem, ahem...".

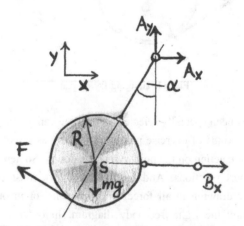

Fig. 12: Free-body diagram of the hanging frying pan

Figure 12 shows the free-body diagram of the system. Here we really do have to isolate the pan from its entire surroundings. This means cutting away everything around it (360°) and substituting the corresponding forces. Make sure you do not forget the weight force, which in this case should act exactly from the middle of the burger torture device (the handle is massless). Again we arrive at three unknown values: A_x, A_y and B_x and therefore need 3 equations once again. But what about B_y ? (watch out!). We already know, of course, that a rope can transmit only tensile forces in its own direction. So the direction of force at support B is already clear. In situations like this you have to do a real Columbo number every now and then, but after all, we're academicians, or at least want to be (or have to be? "...think of your future, my child!!!"... *"Yes... mom!"*).

Hence – the force of reaction B can only act in the direction of the rope (thus: B_x). This case comes up quite a lot and should be something that you should by all means have in your "little bag of tricks". We then resolve the reactions at the two-force support A into their respective coordinate directions x and y. Again we write out our three summation equations for two-dimensional statics (you should not only read the following equations, but follow along by writing them down with paper and pencil):
for the x direction (equation I):

$$\Sigma F_x = 0 = B_x + A_x - F \cos\alpha. \qquad (I)$$

"– F cosα" is the element of applied force F acting in the x direction, the so-called "x component of force F" (please note its negative sign!!!).

The question often arises: sine or cosine? Here it is helpful to see what happens when $\alpha = 0°$. When $\alpha = 0°$, F pulls completely in the x direction and must therefore enter completely into the "x equation (I)". And since the cosine of $\alpha = 0°$ is one, the cosine must be used here. The angle α of force F follows from the geometric consideration that the line of action of F is perpendicular to the middle line of the pan, which itself is inclined at an angle α to the wall and thus to the y axis. These are things you'll just have to practice until you get the hang of it. After about 50 exercises you'll be able to make these kinds of connections immediately (according to Dr. Hinrichs, he himself only needed about 40 exercises).

So, on we go... for the y direction (equation II):

$$\Sigma\, F_y = 0 = A_y + F \sin\alpha - mg\,, \qquad\qquad (II)$$

where "F sinα" is the y component of force F and "mg" describes the weight force which always results from the product of the mass and the acceleration due to gravity g (ca. 10 m/s^2). (Dr. Hinrichs can only calculate this using at least two decimal places, i.e. g = 9.81 m/s^2.) In the unit of acceleration, time t is squared! ...? This is no cause for alarm, much less for anything unnatural. Even in the field of mechanics, the totally inadequate, but usually arrogant "scientist" can conceive of time only in terms of a one-dimensional entity (although according to Dr. Romberg, it, like space, likewise has a number of easily comprehended dimensions). The expression "s^2" is in the denominator of the unit simply because here one is dealing with a change in velocity per second, or in other words, acceleration. You might say: The unit of acceleration a, which plays a major role in dynamic systems (see Chapter III), is expressed as meters per second... per second, thus m/s2... it's really quite simple, isn't it?

But now back to our equation (III) (summation of moments). What would be our best choice for the point of reference? F is known... mg is known... hmm... the most unknown forces are excluded at point A, plus we can then let the entire force F "rotate", because it just happens to be perpendicular to its lever arm (otherwise we would have to calculate force components here as well). So let's choose A as the reference point. Equation (III) is therefore:

$$\Sigma \, M_z^{(A)} = 0 = mg(L+R)\sin\alpha \; + B_x(L+R)\cos\alpha \; - F(L+2R). \text{ (III)}$$

"(L+R)sinα" is the perpendicular lever arm of force mg around point A, "(L+R)cosα" is the perpendicular lever arm of force B_x . Knowing from above that L=2R, we can calculate the unknown force B_x from the moment equation (III) and from equation II we immediately get A_y:

$$B_x = F \, 4/(3\cos\alpha) - mg \, \tan\alpha,$$
$$A_y = mg - F \, \sin\alpha \, .$$

By putting Bx into equation (I), we then get, after simple transformation:

$$A_x = (\cos\alpha - 4/(3\cos\alpha))F \; + mg \, \tan\alpha.$$

In order to put a tricky spin on these kind of exercises, nasty little questions are often asked, such as: "What is the minimum magnitude of force F needed to take the slack out of rope S?" The panic released by such questions – slowly creeping up ice-cold from the base of your spine, fighting its way to your neck and, despite its ghastly chill, forcing sweat out of every pore – can be smashed by a simple yet highly effective trick (in short: "It takes 'the P' out of your face!"). So when faced with such exercises or test questions, proceed as follows: After taking a deep breath, twist the corner of your mouth in a slightly contemptuous grin aimed in the direction of your nearest despairing classmate, then ask yourself the following two questions.

1) Can I equate anything here?
2) Can I equate anything here to zero?

In over 95% of all cases you'll be able to answer one of these two questions with "Yep, sure can!" For the remaining 5% or so, go on to the next question and get ready to pick up your loser ticket.

Let's try out this trick on the above problem about the slack rope and the magnitude of force resulting from it. Can you equate anything for a slack rope? Can you equate anything to zero for a slack rope? Yep, sure can! Namely, the cable force! When the rope is slack, it transmits no force to support B. This means that we can equate force B_x to zero in all equations, provided that it arises in the first place. Hence, force F is obtained directly from the momentum equilibrium (Equation III) for a rope at the point where it slackens as

$$F = 3/4 \text{ mg sin}\alpha \ .$$

So the answer to the above question is that force F must possess at least 3/4 mg sinα to keep the rope taut.

1.5 Determinedly Statically Determinate... Right?

In the examples we have been working with up to now, we have always had the exact number of equations at our disposal to cover the unknown forces in the problem to be solved. Unfortunately, this is not always the case, for there also happen to be systems which cannot be dealt with on the basis of equilibrium conditions alone. This can happen, for example, if you have more unknown forces than available equations. For exam questions you can usually assume that "the calculation will tally". It is said that there are even people who coolly calculate ignorance for the rest of the cases.[5] But particularly in practical applications you will often have a few supports too many, something referred to as a statically indeterminate(redundant, s.u.) system. These are the sort of things we would now like to take a closer look at.

---

[5] At this point the obligatory protest is made by Dr. Hinrichs.

What does "statically determinate" mean? Memorize the following sentence: A system whose support reactions can be determined solely from the conditions for static equilibrium is considered to be statically determinate. Conversely, it follows that a system in which the conditions for static equilibrium are *not* sufficient for determining the unknown support reactions is statically indeterminate.

So let's take a look at the following system (Figure 13):

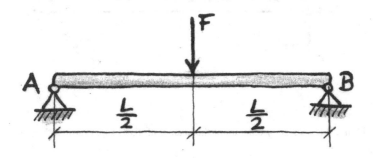

Figure 13: Beam with fixed supports at both ends

Figure 13 shows a schematic drawing of a rigid, massless beam which lies on two fixed supports. What kind of support reactions can we expect here? Stop!!! What is the first thing you do in statics before formulating any other thoughts? We must first isolate the system, drawing in all forces and moments correctly, or in other words, we first have to draw the right free-body diagram. The first thing we do in determining the support reactions is to draw a free-body diagram. At the start it is very important to draw a proper free-body diagram. Before we think of anything else we first have to draw a free-body diagram.

We first draw an intelligent free-body diagram before we start calculating the support reactions. At the start of every statics problem you first draw a free-body diagram. The first thing we do is isolate the system! We draw a free-body diagram. Even before we scribble down the first equations! The first thing we do is draw a free-body diagram. At the start of every statics problem we

31

therefore draw a correct free-body diagram. First we isolate the system by cutting it free. Isolate – the most, the very most (yawn) important thing of all... We really like free-body diagrams. So what we are doing first is looking for the appropriate free-body diagram. The corresponding free-body diagram is illustrated in Figure 14.

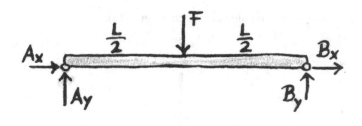

Figure 14: Free-body diagram of the beam with fixed supports at both ends

You can see at once that there are four unknown forces here, while for the two-dimensional world only our three famous equations are available to us. From the force equilibrium in the y direction and the summation of moments around an arbitrary point (one of the supports is the best choice) we first obtain

$$A_y = B_y = 1/2 \, F \, .$$

By now everyone should be able to work that out on their own. If not, close the book and start reading from the beginning tomorrow! But what about horizontal forces A_x and B_x? All of the Nobel laureates[6] in mathematics in the world together could not calculate these two forces without further information. The only thing we know from the force equilibrium in the x direction is that the two forces must cancel out:

$$A_x = - B_x \, .$$

---

[6] Important note by the editor: "There is no such thing as a Nobel Prize for mathematics!"

These two forces may be of any magnitude without altering anything in the mechanical character of this rigid beam. To anyone asking about where these unknown forces are supposed to come from, it should be pointed out that this beam could have been squeezed, even with considerable force, between the supports without being deformed at all. This can generate some pretty mean forces in the horizontal direction (x). The system is statically indeterminate, and even in the most literal sense of the word it is probably more indeterminate than next week's lottery numbers. Mechanics call this system statically redundant. You can also say: The system is jammed. This is exactly why such beams, rods, shafts, bridges, etc. in practice are always, as a matter of principle, and everywhere – mechanical engineers know this – provided with a fixed support and a roller support. This arrangement gives you tension-free bearing and the horizontal force is zero as long as no external forces come into play. And even if they do, the fixed support just absorbs them.

The bright loser will certainly be able to imagine what a statically underdeterminate system might look like here. The slightest external force component would send the beam in Figure 13 with two roller supports for a ride, since any support reaction is lacking. A statically redundant system, on the other hand, means that there are simply too many support reactions present, or too few equations. This means that we need more equations to calculate a statically redundant system.

Additional equations can be obtained from so-called *intermediate conditions*. These are based on the following idea: It is basically permissible to isolate even individual members of a body or system. For example, we can cut away a corner of any body and draw in at the resulting interface – very "quickly", before the corner "notices" anything – the corresponding (unknown) forces and moments, which otherwise hold the body together exactly at this point. But this step does not do us any good unless it provides us with more equations than the new and uninvited unknown forces and moments revealed at the interface. But how is that done? As a rule when working in two dimensions, whenever we cut through a body at any point, we have to draw in three reactions: two force components (horizontal and vertical), and a moment, which hold the body together exactly at this point before anything has been cut off. But by doing so, we also obtain only three new equations (we'll see how and why in just a minute).

But now comes the trick:

If we isolate exactly at a joint, we only need to draw in two force components, since a moment of reaction can never get a grip on an ideal frictionless joint.

Let's turn to the following example:

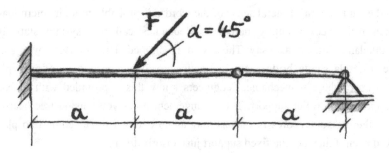

Figure 15: Beam with intermediate joint

In the illustrated constrained and massless beam with an *intermediate joint* (see Fig. 15), the other end is not free but supported by a roller support.

You can always spot a real mechanics person by their instinctive reaction at the sight of such an ideal structure: They grab (slobber, tremble) paper and pencil – or a notebook and the latest version of Corel – and draw an "intelligent" free-body diagram. Remember, we must first isolate the system and correctly draw in all forces and moments, or in other words, we first draw the correct free-body diagram. For determining the support reactions we must first and foremost draw a free-body diagram. The most important thing at the very start is to draw a correct free-body diagram. Before we even start to consider the problem further, we first draw a free-body diagram. We first draw an intelligent free-body diagram before we start calculating the support reactions. At the start of every statics problem you first draw a free-body diagram. The first thing we do is to isolate the system! We draw a free-body diagram. Even before we have really understood the problem in the first place!

We start by drawing a free-body diagram. At the start of any statics problem we draw a great free-body diagram. First we cut free and isolate.

Cutting free – say goodbye to Loserville! The corresponding free-body diagram of the massless beam looks like this:

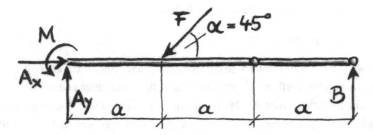

Figure 16: Free-body diagram of beam with intermediate joint

Here we now have the case where too many forces are acting without any justification... we still don't know how roller support B and fixed-end share the load between them. The system appears to be statically redundant. We have four unknown reactions but only three equations.

Now for the trick with the intermediate condition and the bold cut right through the joint. The result is two individual free-body diagrams, with the intermediate reactions occurring between the two free-body diagrams always having to fulfill the action = reaction axiom. This means that the intermediate forces for both free-body diagrams are equal in magnitude but labeled with the opposite signs.

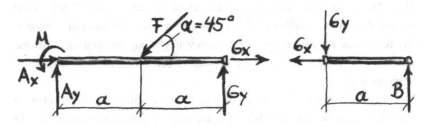

Figure 17: Cut through the intermediate joint

The beauty of this is that we now have three equations for each of the two sub-systems.

From the summation of moments about the intermediate support G of the sub-system on the right, it immediately follows that:

$$B = 0 .$$

Now we can relax and go back to our over-all system, since there are only three unknown reactions left, which we can bring to light with three equations!

The joint that saved the day is known as a *Gerber joint*. The entire beam is called a *Gerber beam*. The beam, which at first appeared to be statically redundant, has been "made statically determinate" by the joint, i.e. we have adapted the system. (There are also other ways of handling statically redundant systems, see Chapter 2)

This system (Gerber beam) was just meant to demonstrate how helpful intermediate conditions can be to you. Of course, in practice you are not allowed to start drawing in a joint somewhere just to better cope with a statically indeterminate system. But when dealing with a (rather complicated) system, if you're a sharp looker (wow!) you'll usually find a convenient place somewhere to make a surgical free-body cut that gives you an elegant intermediate condition, thus providing you with all sorts of new independent equations to your heart's delight. At this point, however, we would like to limit ourselves to the *detection* of statically indeterminate systems.

Since the pragmatic and technically "gifted" engineer (aren't they all?) must always have a definition or formula at hand for testing a hypothesis (Dr. Hinrichs knows this), there is also a kind of "counting rhyme" for testing statical determinacy. Although you really do not need this formula – since in the few cases where statical indeterminacy might pop up, you can also use common sense for a change – we decided to present it here for the sake of completeness.

Hence, a system of equations can only be solved if the number of unknowns matches the number of equations. A (necessary) condition for statical determinacy in two dimensions meets the following equation:

$$D = 0 = \Sigma a + \Sigma z - 3n ,$$

where:

D = defect

a = number of reactions (valency) per support

z = number of reactions per intermediate condition

n = number of members separated by joints (intermediate
      supports).

Unfortunately, this condition has a small catch: It is merely a *necessary*[7], but not a *sufficient*[8] condition.

This condition must be met in a statically determinate system; but you should not conclude from this condition alone that the system is actually statically determinate. OK, once again. The above equation is a necessary condition, i.e. for a statically determinate system the defect D is always zero. But under certain unfavorable circumstances it can also be zero for a statically indeterminate system too, so watch out! But a system where D ≠ 0 is inevitably statically indeterminate.

_____

[7]Incidentally, Dr. Hinrichs is not necessary.
[8]Please note: Dr. Romberg often demonstrates he is not sufficient!

For three-dimensional systems, the "3" is substituted by a "6". (The possibility of rods in space rotating on their axes does lead to statical indeterminacy, which you shouldn't take too seriously, since such systems can be calculated anyway.) Here are two tips for setting up the above equation:

- The number z of intermediate conditions of a joint connected to n members is $2(n - 1)$ (e.g. Gerber beam: $n = 2$, $z = 2$)!

- Systems that exhibit a maximum of one force acting at each and every one of its supports are always statically indeterminate.

Now is a good opportunity to go back over the examples covered in this section using our "counting rhyme"... have fun!

Despite the energetic protest of Dr. Hinrichs, it is highly recommended that you check any system suspected of being statically indeterminate with the unscientific "wobble method". This is quite simple: Pretending that the system being analyzed really exists as rigid structural members, we grab it with an imaginary hand and shake... if there is anything that somehow might be able to wobble, then statical indeterminacy is present. Dr. Hinrichs objects that you should not take it to heart if every now and then something is not really stable, this happens a lot and is perfectly normal! (?) If the system is still stable after giving it a shake, it can still be statically redundant. In this case, where the structural member remains solidly in position, we'll take another careful look and consider whether anything might be jammed (statically redundant). Hopefully, the illustrated examples will make this clear. In the example in Figure 18, at the very top left, defect $D = 0$ ($a = 3$, $n = 1$, $z = 0$) but the system is nevertheless statically indeterminate. But this is only because it "jams" and "wobbles" simultaneously. If that is the case, defect D may cancel out but we still have a statically indeterminate system. This correlation[9] will hopefully be verified by the other examples in Figure 18.

---

[9] Dr. Romberg notes that this is the first time this correlation has been mentioned as formulated here and he would therefore like to claim the "Defect-free Coinciding Wobble Clip (DCWC)" as his own discovery.

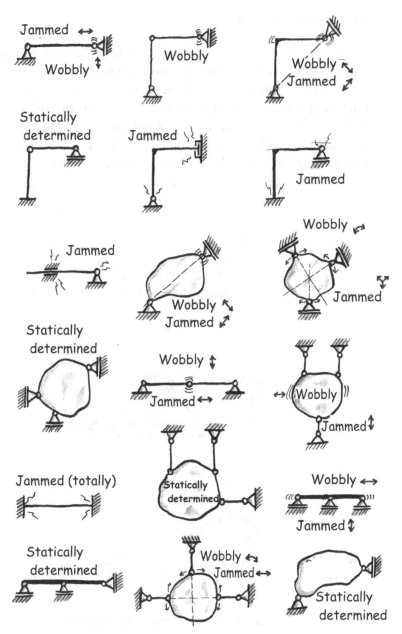

Figure 18: Some examples of statically (in)determinate systems

1.6 Distributed Loads

Forces do not act necessarily at a single point, as has been demonstrated up to now. If we imagine, for instance, someone lying lazily stretched out on a plank bed (see Figure 19 a, b, c), the bed will be subjected to different loads along practically its entire length, depending on the anatomy of the person. Summing up the resultant of this distributed load generally requires mathematical integration. Here the distributed load q is specified as a function q(x) which describes a force per unit length (newtons per meter).

a)

b)

c)

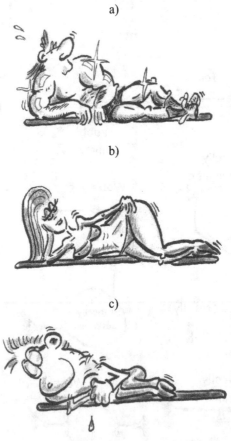

Figure 19 a, b, c: Differently loaded plank beds

Usually, however, you can also employ another very simple way of looking at the center of gravity. If you know the "mid-point" (center of gravity) of a distributed load, then you just have to imagine the "sum force" (integrated by length) as acting precisely at this point. The support reactions are calculated as usual. But where does the point of application for this force lie? First of all, let us consider how to calculate the center of gravity.

1.7 Center of Gravity

In the following we will assume homogenous, rigid bodies. A homogenous body has the same (physical) properties at every point. With respect to the center of gravity, this means that a homogenous body exhibits uniform density so that the center of gravity ultimately depends on the shape of the body alone. The center of gravity is then identical to the geometric mid-point of the body's volume. When working with two dimensions, which is the case in 90% of all problems, we're simply dealing with the sum mid-point of all surfaces. So once again, let's make a few more or less intelligent assumptions so that we can at least get some kind of grip on a problem. Let's take a close look at the figures in Figure 19 (all three, please):

In the first case (Fig. 19a) we have a classic muscle freak, regular patron of your local fitness studio (uggghh!). There is definitely no homogeneity here, since in reality his head is of significantly less density ($\rho \to 0$) than his biceps. But we shall make the bold assumption that here we are also dealing with a homogenous body.

The body in Fig. 19b ("wow!" ← quote from Dr. Hinrichs) is homogenous. Here, too, there is an unequal distribution of mass across the plank bed (length L). In the case of the electrical engineer in Fig. 19, there is also a significant imbalance.

But how exactly is the plank bed loaded? Where is the point at which the resultant weight force acts, which we need to know in order to calculate the support reactions? The magic word here is: modeling. Machines or building structures are too complicated to calculate their center of gravity exactly. But you can approximate reality fairly well through clever modeling.

The obvious thing to do here is to split the body into smaller shapes whose "centers of gravity" are known. We will view the three figures as being two-dimensional, modeling them by using squares, circles and triangles.

Figure 20 shows the so-called "equivalent system" for each body, each one comprising just these simple geometric forms (Dr. Hinrichs has insisted on the finer details for Body b).

a)

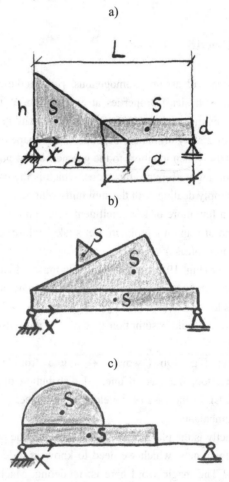

b)

c)

Figure 20 a, b, c: Equivalent systems for the bodies

42

The important thing now is that we know the centers of gravity of the individual parts. The centers of gravity of circles, squares and rectangles are known to every head loser or loose head. For triangles and semicircles it's a little different. The center of gravity of a homogenous triangular plate of uniform density is the point at which the median lines intersect. If height H is defined from an arbitrary base of the triangle, then it holds true for the center of gravity x_s that:

$$x_s = H/3.$$

The overall center of gravity, or centroid[10], is then formed from the weighted average of all distances of the partial centers of gravity:

$$x_s = (\Sigma\, x_{si}\, A_i) / (\Sigma\, A_i).$$

For our bodybuilding freak from Fig. 19a, applying this first for the summation of the weighted partial centers of gravity in the x direction, we get:

$$\Sigma\, x_{si}\, A_i = b/3\ bh/2 + (L - a/2)\ ad\ ....$$

But this leaves us with one small problem. Since the two geometric shapes overlap in the middle, we would be exaggerating the forces in the genital region of this typical mechanical engineer. The small shaded triangle (see the equivalent system) appears twice in our calculation, so we'll have to subtract it once from the summation of the weighted partial centers of gravity. The correct calculation would then look like this:

$$\Sigma\, x_{si}\, A_i = b/3\ bh/2 + (L - a/2)\ ad - ((L - a)+(b - (L - a))/3)\ (d\ (b -(L - a))/2),$$

with the term in italics representing the product obtained from the center-of-gravity coordinate of the small triangle in the middle and its area. Try working through this once more on your own! Now all we have to do is divide this expression by the overall area – and you got it!!!. This is easier to see if we set up the equivalent system differently (see. Fig. 21). However, here you should obtain the same result for the center-of-gravity coordinate x_s... give it a try!!!

---

[10]Incidentally, for a semicircle, the distance from its center of gravity to the sectional edge is $y_s=4R/3\pi$.

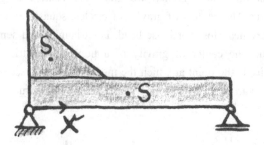

Figure 21: Alternate equivalent system for bodybuilding freak

The centers of gravity for the other two bodies can be determined the same way. For a three-dimensional model you just make the same calculation for each of the three coordinates.

1.8 3-D Statics

For three dimensional systems, you either have to determine the summation of forces and moments for each coordinate direction, or resort to vector calculation. For the first method you will need a good sense of three-dimensional structures, as well as little bit of time and intuition (so that nothing gets left out). For the second method all you need is a good dose of engineering pragmatism (despite the fact that engineers, who are well-known leaders in aesthetic taste, consider this the more "elegant" method). So let's look at the following simple example:

A certified engineer (mechanical) is interested in developing the "idea" of having the visor of the familiar Rapper cap supported by a thread (see Fig. 22). The problem forcing itself upon the engineer is how to size the thread, thus presenting the question concerning the force of the thread F_S at the given weight of the visor G.

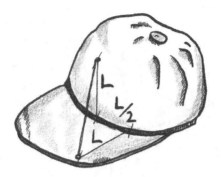

Figure 22: Visor cap with thread

Again the first thing we need to do is to find the appropriate equivalent system. Departing from reality, let's consider the visor as a two-force homogeneous plate whose weight force G acts upon the center of gravity S. After a series of talks with industrial designers flown in from Milan (the buffet was OK, but oh how they skimped on the carpaccio! Porca la miseria!) we came to the conclusion that the suspension point A was placed in a "corner" of the visor at the origin of the coordinate system (see Fig. 23).

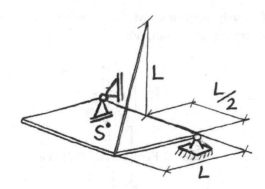

Figure 23: Equivalent visor system

What do we do next? We isolate! Figure 24 shows the free-body diagram of our equivalent system.

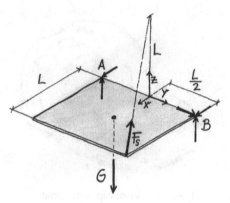

Figure 24: Free-body diagram of the visor

Instead of calculating the equilibrium equations for each coordinate direction individually, we'll apply the much more efficient vector method of calculation here, assuming, of course, that the reader is already familiar with vector calculation. (Since three-dimensional statics in general, and even in particular, are rarely asked in test problems, you can also skip over this section, uh, oh but now I hear angry protests coming from Dr. Hinrichs... Chin up, Doc!)

A look at the free-body diagram shows the following force vectors (in the following, vectors are printed in **boldface**):

weight force:
$$\mathbf{G} = \begin{bmatrix} 0 \\ 0 \\ -G \end{bmatrix},$$

cable force:
$$\mathbf{F_S} = \begin{bmatrix} -L \\ -L/2 \\ L \end{bmatrix} 2/(3L) \, F_S,$$

support A:
$$\mathbf{A} = \begin{bmatrix} A_x \\ 0 \\ A_z \end{bmatrix},$$

46

$$\text{support B:} \qquad \mathbf{B} = \begin{bmatrix} B_x \\ B_y \\ B_z \end{bmatrix}.$$

Here, F_S is the magnitude of the yet unknown cable force. The factor in front of this magnitude is the "standard" value of the force vector calculated from its geometry (see Fig. 24). Based on this free-body diagram, the equilibrium conditions expressed as vectors are

for the forces:

$$\Sigma \mathbf{F} = \mathbf{0} = \mathbf{G} + \mathbf{F_S} + \mathbf{B} + \mathbf{A},$$

for the moments:

$$\Sigma \mathbf{M} = \Sigma (\mathbf{r} \times \mathbf{F}) = \mathbf{0}$$

$$\Leftrightarrow \begin{bmatrix} L/2 \\ 0 \\ 0 \end{bmatrix} \times \begin{bmatrix} 0 \\ 0 \\ -G \end{bmatrix}$$

$$+ \begin{bmatrix} L \\ L/2 \\ 0 \end{bmatrix} \times 2/(3L)\, F_S \begin{bmatrix} -L \\ -L/2 \\ L \end{bmatrix} + \begin{bmatrix} 0 \\ -L/2 \\ 0 \end{bmatrix} \times \begin{bmatrix} A_x \\ 0 \\ A_z \end{bmatrix} + \begin{bmatrix} 0 \\ L/2 \\ 0 \end{bmatrix} \times \begin{bmatrix} B_x \\ B_y \\ B_z \end{bmatrix} = \mathbf{0}.$$

The second component of this vector equation leads directly to the cable force

$$F_S = \tfrac{3}{4}\, G \quad.$$

In three dimensions, a moment is formed by the cross product of force vector and lever arm, since these two vectors, of course, do not necessarily need to be orthogonal to one another. At the sight of this vector equation Dr. Hinrichs cannot repress an ecstatic gaze of rapture, while Dr. Romberg apologizes for both.

1.9 Now There's Going to Be Some Friction...

1.9.1 Frictional Forces and Coefficients of Friction

Up to now, everything has certainly been going smoothly. But friction itself is a very important topic (and not only in mechanics). What would the world be like without friction? It's a lot of fun to think this idea through to its logical conclusion. In a world where only interlocking connections were possible, you would even have a pretty hard time just moving forward, for instance.

Friction is practically everywhere. Let's consider the following experiment: If you pull a coffee cup across the table, you will notice a resistance that's always pointed in the direction opposite to the motion. If we now attach a rubber band to the handle, we can observe the acting friction force as a function of the level of coffee in the cup (weight force) by looking at the expansion of the rubber band as we pull the cup lightly and effortlessly across the table (the only other thing which is easier to pull something over on is Dr. Hinrichs himself, but that's another story).

We see that the more coffee there is in the cup, the longer the rubber band stretches at the start of the movement or while the cup slides across the table. And something else can be clearly seen: The rubber band is always stretched the longest right at the start of the movement, i.e. static friction is greater than the sliding friction. Since no other forces act in the direction of pull at constant velocity or at rest, force F_G in the rubber band must correspond exactly to frictional force F_R. The best way to visualize this is – naturally – to make a free-body diagram of the cup being pulled across the table.

We have already established that the tensile force (=frictional force F_R) increases with increasing weight (coffee). If you conduct this experiment under "laboratory" conditions, you will see that the tensile force is directly proportional to the weight force G.

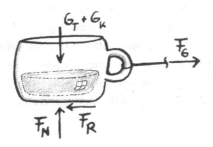

Figure 25: Free-body diagram of coffee cup with rubber band

In our example this corresponds to normal force F_N (thus: $G = F_N$). Being proportional means that the relationship between two variables is constant. It therefore holds that:

$$F_R / F_N = \text{const.}$$

Here we restrict ourselves to the simple Coulomb friction model, which describes two different states: adhesion (velocity v=0) and sliding (v ≠ 0). In the case of sliding, this constant factor of proportionality is referred to as the *coefficient of kinetic (*or *sliding) friction* μ, thus

$$F_R = \mu F_N .$$

Before there is any motion, the frictional force F_R is indeterminate, only fulfilling the unbalanced equation

$$| F_R | \leq \mu_0 F_N ,$$

with μ_0 being the *coefficient of static friction*. When you slowly start pulling the slack rubber band[11] without initiating the sliding process, you can easily imagine that, in terms of any given direction, the resultant of the two forces F_R and F_N will be located within a cone, the so-called *cone of friction*. One-half of the cone angle α_0 of the cone of friction is calculated as

$$\tan\alpha_0 = \mu_0 .$$

---

[11] This does not refer to the drawing on the next page but to the rubber band tied to the coffee cup.

Or to put it more simply: If we tip an incline with a block on it until the block starts to slip at angle α_0, we then obtain $\tan\alpha_0 = \mu_0$. As kinetic friction occurs, the corresponding $\tan\alpha = \mu$ applies, with the resultants of both forces forming something like an envelope of the cone of friction. The important thing to remember here is that the friction force F_R is independent of the support surface and that it depends only on the normal force F_N and the coefficient of friction μ or μ_0.

According to the Coulomb friction model, friction force depends only on the material mating (μ) and the normal force. Consequently, there are no other parameters of influence involved. Even good old Leonardo conducted experiments on this and came up with precisely the same conclusion. Figure 26 shows da Vinci's original drawings of his friction experiments.

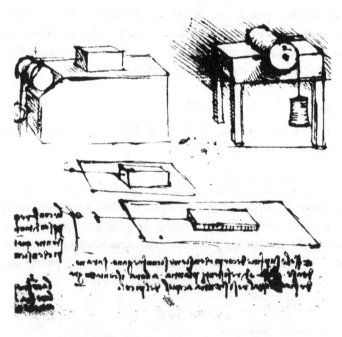

Figure 26: Da Vinci's friction experiments [16]

In test questions involving problems of friction, the main thing to know at the start is that a friction force must be drawn in somewhere in the free-body diagram. But here you can use a really simple engineer's trick: Wherever in the formulated problem a μ or μ_0 is drawn in at the interface between the object to be isolated and its surroundings, the thing to do is draw in the friction force *(but after making the free-body diagram!!!)*. You then have one more unknown... but – watch out! – you also have one more equation that needs to be satisfied, namely:

$$F_R = \mu_{(0)}F_N \; ,$$

where up to the point at which motion occurs this usually involves the maximum possible friction force, thus turning the unbalanced equation into a legitimate equation.

Let's examine the following example:

Figure 27: Tug-of-war on ice

We already met Sir Isaac Newton (overall mass m) at the start. Now the iron pumper (overall mass M) from the section on distributed loads (see Fig. 19) has challenged our Sir Isaac to a tug-of-war contest on ice. For this, let's consider Figure 27. The coefficient of static friction between the ice and the shoe soles is μ_{01} for Newton's ornate rococo galoshes, and μ_{02} for the high-quality maximum anti-skid Adidas tennis shoes of his opponent.

Here Newton employs a trick to outfox his adversary. In order to raise his friction force, he straps on a backpack. The question now is: How heavy must the backpack be in order that he can start pulling his opponent across the ice? Or to put it differently: At what backpack mass m_R does the friction force F_{R2} on Newton's shoe soles reach precisely the value of F_{R1} (Adidas soles)?

To solve this problem, we isolate the free-body diagram! We must first isolate the system, drawing in all forces correctly, or in other words, we first have to draw the correct free-body diagram. The first thing we do in determining the support reactions is to draw a free-body diagram. At the start it is very important to draw a proper free-body diagram. Before we think of anything else we first have to draw a free-body diagram. We first draw an intelligent free-body diagram before we start calculating the support reactions.

THE NEW EXTREME - KICK FROM USA:
FREECUTTING...

At the start of every statics problem you first draw a free-body diagram. The first thing we do is isolate the system! We draw a free-body diagram. Even before we have even scratched our head in desperation! The first thing we do is draw a free-body diagram. At the start of every statics problem we therefore draw a correct free-body diagram. First we isolate the system by cutting it free.

Cut it free, the only way to be.

Figure 28: Free-body diagram of the two warriors on ice.

We therefore determine:

$$\Sigma F_x = 0 = F_{R1} - F_{R2}$$

$$\Leftrightarrow \quad F_{R1} = F_{R2}$$

$$\Leftrightarrow \quad \mu_{01} F_{N1} = \mu_{02} F_{N2}.$$

From the summation of forces in the y direction it then follows:

$$\mu_{01} Mg = \mu_{02} (m + mR)g.$$

This gives the mass of the backpack as:

$$m_R = (\mu_{01}/\mu_{02})M - m.$$

That's about all for this question – but now on to another source of friction.

1.9.2 Rope Friction

One special case, which nevertheless is frequently encountered in practical applications, is the friction between a rope and a pulley. Here, too, a distinction is made between the states of static and kinetic friction, although for the acting

54

forces it doesn't matter whether the rope or the pulley moves in the case of kinetic friction. Let's first consider the case of static friction, i.e. where no relative motion occurs between rope and pulley. A pulley is shown in the following Figure. It is driven, held fast or braked by a crank (mechanics say that the action of a moment is present in an instantaneous center of rotation).

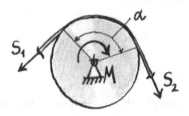

It is quite easy to imagine that the two rope tensions are not just deflected here but instead, depending on the static friction, that a certain ratio of the two tensions S_1 and S_2 is present. For example, if you glue the rope to the pulley, one of the two ropes can go completely slack while the other one is about to snap. If static friction is present, you can get the rope to glide in one direction or the other after a certain ratio is reached. This ratio can be derived from the free-body diagram of an infinitesimal, really, really tiny section of the rope on the pulley surface. ~~The interested reader The diligent student~~ Hot-shots will be able to calculate this relatively short and simple derivation on their own or find it in any "good" mechanics textbook. The interval of the tension ratio at the point of impending sliding, depending on the direction of net tension, leads to

$$e^{(-\mu_0 \alpha)} \leq S_1/S_2 \leq e^{(\mu_0 \alpha)} \quad .$$

To keep from getting confused, it is recommended to note, depending on the direction of the acting drive, braking or retaining moment, which of the two tensile forces is greater. Analogous to the explanations given for static friction, the following can be applied to kinetic friction:

$$S_1 = S_2 \, e^{(\mu \alpha)} ,$$

where here you should "think along" with respect to the magnitude and direction of forces.

"Yeah, great", says the interested loser, "but which of the two limits is decisive?" In order to answer this fair question, we shall now – with Dr. Hinrichs leading us safely by the hand – take a little stroll through the wonderful world of exponential functions. These beautiful yet dangerous entities have namely the following properties:

$$e^x < 1 \text{ for } x < 0,$$

$$e^x = 1 \text{ for } x = 0,$$

$$e^x > 1 \text{ for } x > 0.$$

If you determine by means of a driving, braking or retaining moment which of the two tensions is greater, you will see whether the quotient S_1/S_2 is less than 1 (then the left limit is relevant) or greater than 1 (then the right limit is relevant).

1.10 Trusses

Now we come to the so-called "in your sleep" problems, which bring joy to the hearts of those who are still in the loser rankings. With trusses you can earn a lot of points. As already mentioned above, a truss is made up of rods, which can transmit forces only at their ends and in their longitudinal direction. These rods are connected at their ends exclusively by ideal joints (where no moment is involved). External forces and supports can therefore only arise at these joints. Such a truss does not exist at all in reality, but that does not matter to us right now. The relative error arising for welded or riveted joints in actual gridwork or trelliswork is said to be about only 5%. In the following we shall assume statical determinacy

When calculating trusses you can either take a very pragmatic approach (using reliable but tedious and yawn-evoking methods), or you can use winner tricks which lead you straight to the goal. When working with trusses you also have to construct a free-body diagram (Dr. Hinrichs likes to refer to the isolation of a free-body as the "Divine Method").

Let us recall: You are allowed to cut through, or free, a (non-living) body from its surroundings at any point you like. But you have to make absolutely sure that you draw in all stress values – and this means *all* forces and moments – which occur at the plane (or edge) of the cut. Once you have grasped that, your loser days are behind you!!!

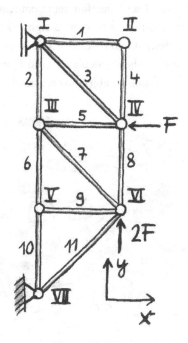

Figure 29: Truss

Let's take the example illustrated above (Fig. 29) and treat it in two ways. Here we are looking for force S_7 in rod 7 (see Fig. 29). At the start of every truss problem the first thing to do is calculate the support forces:

This gives us (after making the correct free-body diagram) and $F_{joint\ I} = A$, $F_{joint\ VII} = B$:

$$A = 4/3F, \ B_x = -1/3F, \ B_y = -2F \ .$$

1.10.1 Slowly But Surely (Method of Joints)

In this reliable method we isolate the free-body diagram joint by joint and, keeping Newton's Third Law in mind (*action = reaction*), draw in all forces, gradually moving hand over hand to the rod whose inner force "interests" us. Here we observe the convention that at first all rod forces are indicated as directed away from the rod. These therefore represent tensile forces (pulling at the joints and the rod), resulting in a positive sign (+) in the calculation. In keeping with this convention, compressive forces are calculated with a negative sign (−), i.e. after we have calculated a rod force and the result has a negative sign, the rod in question is subject to compression.

To make this calculation easier, we can take advantage of a cool trick here as well. Consider the free-body diagram of joint II (see Fig. 30):

Due to their orthogonal relationship (being perpendicular to one another) the forces S_1 and S_4 will never be able to cancel each other out. But joint II (see Fig. 29) is at rest, isn't it!? Nevertheless, the equilibrium of forces in both directions for a body at rest is not fulfilled here... unless: Both forces are zero (0), not present, the rods are "empty" and not necessary at all. In this case we refer to them as being zero-force rods, which can save you a lot of time, in contrast to some colleagues who are a big zero, such as Dr. Romberg from time to time.

An *unloaded isolated corner* (such as joint II in Fig. 29, for instance) or even an unloaded *T-piece* (for example joint V) always indicates zero-force members. Even if member 9 happened to be joined to member V at an oblique angle (oblique T-piece), member 9 would still be a zero-force member because members 6 and 10 cannot compensate the x-component of any force in member 9, for even their forces act only along the direction of the members.

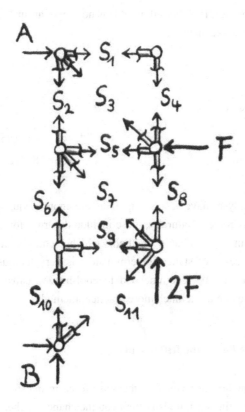

Figure 30: Isolated joints (for joint numbers, see Figure 29)

59

If we take a close look, we'll see that this trick can indeed be applied to member 9: The truss forces S_6 and S_{10} have no components in the "member 9 direction", and can thus not compensate for S_9. The force equilibrium for joint V in the y direction can only be achieved if truss force S_9 disappears. Thus, member 9 is also a zero-force member, which we can remove.

The equations for joint I are:

x direction:	$S_3/\sqrt{2} = -A = -4/3F,$
y direction:	$S_2 = -S_3/\sqrt{2}$.

The equations for joint III (after all, joint II is no longer around) can be formulated in the same manner:

x direction:	$S_5 = -S_7/\sqrt{2},$
y direction:	$S_2 = S_6 + S_7/\sqrt{2}$.

The same applies for joint IV:

x direction:	$S_5 = -S_3/\sqrt{2},$
y direction:	$S_8 = S_3/\sqrt{2}$,
etc... etc...	

After all joint analyses have been set up, there are enough equations on hand to calculate all truss forces. Naturally, these include the truss force S_7 that we're looking for. But this reliable method, which achieves its goal pretty determinably in cases of statical determinancy, is very tedious and in exam situations it can only be recommended to cool-headed calculating whizzes. This is where you can use a much more practical method :

1.10.2 Get to the Point: the Ritter-Cut

The Ritter-cut method lets you find the truss force in question with a single, skillful blow (isolating cut). It also offers you the chance to check critical joints quite quickly.

Furthermore, this method allows you to "eliminate" those "non-essential" parts right away, thus putting you closer to your target joint from the very start. Again we take advantage of the wonderful fact that we, like hobby surgeons the world over, can make a cut anywhere. Which even means right through the entire system.

That would look something like this:

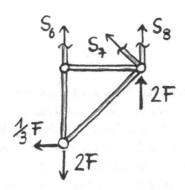

Figure 31: Truss with sectional cut

With criminal acuity (Dr. Hinrichs, go fetch the getaway car!) we first form the summation of moments at the lower remaining sub-system around joint VI, for this then gives us only *one* unknown to deal with. The moment summation results in

$$S_6 = 5/3F.$$

Now all we have to do is make the force summation in the x direction, which immediately gives us:

$$S_7 = -\sqrt{2}/3\ F.$$

Since this trick lets you solve this problem in just a few minutes, you have plenty of time to get up in the middle of an exam and leisurely go for some coffee while the others are sweating it out. It's cool! However, when making the cut, you must always cut through three members only, but these may not be all fixed to the same joint. You'll get the hang of it after some practice!

The Ritter-cut is superbly suited for determining *internal* forces and moments as well. Whenever we isolate a piece from a body we have to supply all the forces and moments in place of this isolated piece so that nothing changes in the original system (yawn). So why do we need internal forces? Quite simple: Internal forces and moments in elastic bodies result in all sorts of deformations (and possible damage). We shall now take a look at internal forces and moments, so-called internal forces (and moments) – but still using rigid bodies.

1.11 Stress Factors

Let's recall the cantilever (the beam fixed at one end) that projects out into space and whose support reactions for a special case of loading we calculated a few pages back. The cantilever in Figure 32 is also considered to be massless at first. We can observe a force $\sqrt{2}\,F$ and a moment M*, which exert a load on the beam, causing *not only* support reactions at the fixed support. There are also forces (and moments) acting along the beam. This is quite easy to imagine if you try holding a safe (mass m) or any other appropriately heavy object (F = mg) in your hand with your arm outstretched. After a while, it is *not only* your shoulder (support) that hurts!

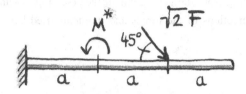

Figure 32: Cantilever loaded by force and moment

How is it possible to determine the so-called inner forces and moments, also known as *stresses*? Well, what do we do when we are supposed to analyze the action of forces and/or moments? We rejoice and draw a free-body diagram (drool!! slobber!! scribble!!) and soon half the battle is already won. So let's first calculate the support reactions.

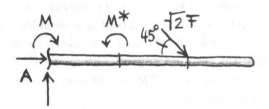

Fig. 33: Free-body diagram of the loaded cantilever

After drawing the correct free-body diagram, the following support reactions emerge:

$$A_x = -F, \; A_y = F, \; M = M^* - 2Fa .$$

Now we determine the stresses, first for the beam section having $0 < x < a$ (thus resolving the system into subsystems whose borders represent the points of external forces and moments). To do so, we simply cut through the left part of the beam somewhere in this section and at the isolated part "very quickly" draw in the forces and moments which in reality hold the beam together there. In this way, the mechanical system does not "notice" anything about the cut, since all reactions are still present. Here, however, we have to observe the ~~Gentlemen's~~ Mechanics' Agreement that stipulates the direction of the reactions projecting from the gaping cut surface (see Fig. 34 and commit it to memory! The directions of the forces and moments used here are henceforth *binding*).

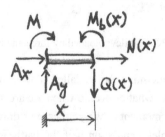

Figure 34: Free-body diagram of the left part of the cantilever (positive section)

A beam is acted upon by a normal force N (tensile or compressive stress), a transverse force Q (shear forces), and a bending moment M_b. *These quantities can now be viewed as a function of a particular location (i.e. of x)* (much in the same way that Dr. Hinrichs is a function of certain locales).
Given that $\Sigma F_x = 0$, $\Sigma F_y = 0$ and $\Sigma M = 0$, the stress values for the left part of the beam can be calculated as:

Normal force: $N(x) = - A_x = F,$

Shear force: $Q(x) = A_y = F,$

Bending moment: $M_b(x) = A_y x + M = A_y x + M^* - 2Fa = F(x - 2a) + M^*.$

In this case and for this part of the beam, only the bending moment M_b is a function of the location x. For determining the bending moment curve, we form the respective summation of moments about the *instantaneous cross-sectional point*. Now let's have a look at the middle beam section at (a < x < 2a). The

64

corresponding free-body diagram is shown in Fig. 35. The stress values for this range are:

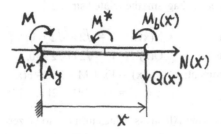

Figure 35: Free-body diagram of the middle beam part (positive section)

Normal force: $N(x) = F,$
Shear force: $Q(x) = F,$
Bending moment: $M_b(x) = Fx + \cancel{M^*} - 2Fa - \cancel{M^*} = F(x - 2a)\,.$

In terms of force, nothing has changed for the two sections, since nothing in terms of force happens in the range $0 < x < 2a$. A force is first introduced at $x = 2a$. But a jump occurs in the bending moment curve $M_b(x)$ at $x = a$, because here is just where the moment M^* is introduced into the beam. The best way to visualize the introduction of a moment is to imagine that someone has just applied a ratchet or a wrench at exactly this point.

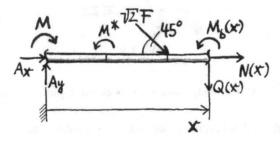

Figure 36: Free-body diagram of the right beam part (positive section)

65

What happens in the right part of the beam? If we take a practical look at this *free end,* it really should be quite clear that no stresses can be generated here anyhow. After all, this right part is not clamped or jammed anywhere... so let's see. First, the free-body diagram: the related stresses:

Normal force: $\quad N(x) \quad = F - \sqrt{2}\,F/\sqrt{2} = 0,$

Shear force: $\quad\ \ Q(x) \quad = F - \sqrt{2}\,F/\sqrt{2} = 0,$

Bending moment: $\quad M_b(x) = Fx + M - M^* - F(x - 2a)$
$$= Fx + M^* - 2Fa. - M^* - F(x - 2a) = 0.$$

Well what do you know! All stress values turn out to be zero (0)! So here there is indeed (what does "indeed" mean anyway?) a jump in the curve of normal force and shear, for the very reason that a shear and a normal force are introduced at $x = 2a$.

But there is a much easier way for us to arrive at the result for the free end, namely by resorting once again to a great trick that saves us a lot of time and calculating effort: Since we are allowed to isolate wherever we wish, we can simply cut off the free end and draw in the section's stress values. But here you have to draw in the section forces the other way around (action = reaction), as was also the case with intermediate conditions (key word: Gerber joint). The common mechanics person also refers to this as the *"negative section"*:

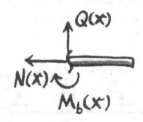

Figure 37: Free-body diagram of the free end (negative section)

So based on the isolated beam end we now just "calculate" the stress values by means of the negative section. With $\Sigma\, F_x = 0$, $\Sigma\, F_y = 0$ and $\Sigma\, M = 0$ the direct result for Fig. 37 is:

Normal force:	$N(x) = 0,$
Shear force:	$Q(x) = 0,$
Bending moment:	$M_b(x) = 0.$

Finding the right free body is something quite wonderful (♥)!

Naturally you can calculate the other parts of the beam proceeding from the negative section. This can be done by anyone who feels like it (Dr. Hinrichs already has pencil and paper in his hand, which is what he always has in his hands unless he happens to be sitting in one of those I'm-really-not-hungry-or-thirsty-at-all-but-it-can-be-something-pretty-expensive-anyway type of bistro, nipping a pretentious, colorful drink and smiling at conceited, stuck-up ladies... and, well... with those of them having no taste sometimes smiling back.).

If a number of supports are present, their reactions are calculated as usual beforehand, with the corresponding values being further treated as external forces. To get the big picture, it is very helpful to plot the stress values across the entire length of the beam. Here, too, there are conventions to be observed, e.g. that positive values point downward. Figure 38 shows the gradients of the three stress values across length x.

Another important tip, especially for oral exams:

With a little practice (Dr. Hinrichs believes that this is possible without any practice), the stresses can be plotted spontaneously with the help of the known threshold values.

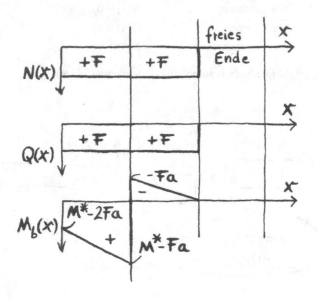

Figure 38: Stress gradients across the loaded cantilever from Fig. 32

The threshold values are given by such forces as support reactions, introduced forces and moments, or free ends. So if you know that the gradients between the introduced forces are constant, and also know that the shear force gradient represents the first derivation of the moment gradient, and that introduced moments cause a jump in moment, then you have been a loser long enough! When working with distributed loads, however, you do have to watch out a little, as shown in the following example. Nevertheless, you have nothing to fear from mechanics if you are consistent, methodical and really cool in applying what you have already learned - and if you set up the right free-body diagram.

We will now apply the method of stress determination to a real cantilever having mass (weight G) with a simple fixed end. The actual case in question involves a one-meter diving board (length L) of the municipal recreational facility complete with a large shower room. Dr. Hinrichs (weight 2G) has been standing for quite some time at a distance b from the fixed end. Despite the encouraging shouts from the thronged crowd below (swimming class of the Delmenhorst Water Fleas) he is still afraid to jump. We first draw the free-body diagram, where we can neglect the fluctuating load of the force exerted by Dr. Hinrichs's trembling knees.:

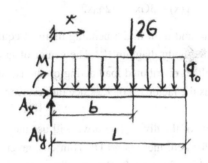

Figure 39: Free-body diagram of a real cantilever (diving board)

The distributed load (weight) $q(x) = q_0$ acts on the entire beam (length L). The resultant weight force is thus $G = Lq_0$. This impinges at centroid $x = x_s$. Along with the additionally exerted force $F = 2G$, which acts at $x = b$, this yields the following support reactions:

$$A_x = 0, \quad A_y = 3\,G, \quad M = G(x_s + 2b) .$$

Now we draw the free-body diagram for the left part (positive section, $x < b$) of the diving board:

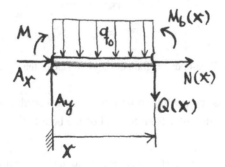

Figure 40: Free-body diagram of the left part of the diving board

69

Here the stresses to the left of Dr. Hinrichs ($x < b$) are:

Normal force: $N(x) = 0$,

Shear force: $Q(x) = A_y - q_0 x$ (distributed load q_0 acting on length x)

                        $= 3G - q_0 x$,

Bending moment: $M_b(x) = 3Gx - 1/2q_0x2$.

Where does the right-hand term in the bending moment equation come from? Take a look at the free-body diagram!!![12] The center of gravity and thus the lever arm of the resultant distributed load acting on the isolated portion lies at $x_s = x/2$. This gives us the factor 1/2 or x^2, since the resultant distributed load is q_0x. Bingo?

For the right side of the diving board we will use the left-hand section to make things easier, for here there is no Dr. Hinrichs, green in the face (from fear) and blue in the lips (from cold). Besides, this is where we don't need to calculate any applicable support reactions. Thus the free-body diagram of the right side:

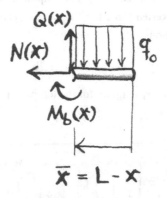

Figure 41: Free-body diagram of the right side of the diving board

First of all, here it is worthwhile to introduce a new coordinate $\bar{x}$ which points in the opposite direction of x from the end of the board. Thus: $\bar{x} = L - x$. This

---

[12] In the case of Dr. Hinrichs we prefer not to cut away his trunks in our free-body diagram.

therefore results in the following stress gradients for the end of the board to the right of Dr. Hinrichs:

Normal force: $\quad N(x) = 0,$

Shear force: $\quad Q(x) = q_0 \bar{x} = q_0(L - x),$

Bending moment: $\quad M_b(x) = -1/2 q_0 \bar{x}2 = -1/2 q_0(L - x)2$.

Naturally, the gradients can be represented in graphic form here as well. Have fun!

Well, enough rigidity – now it's time for some real action, where things bend, pull, hack, wear, tear, shear and squeeze. In other words: "strength of materials".

2. Enough Rigid Thinking: Elastostatics

Here's a little joke about our friends in mathematics:
How does a mathematician catch at least one lion? By driving three posts into the ground, putting a fence around them, placing himself inseide and then defining himself as being on the outside!

Well, now we'll have to dump quite a bit of what we've learned up to this point. This is because one basic assumption from Chapter 1 has to be disregarded entirely. From now on, bodies are no longer rigid, but elastic. We're getting closer to something approaching reality here. Taking an everyday and easily understandable example, let's imagine a pierced nipple, which has a number of weights on it to enhance the arousal factor. Going by Chapter 1, we can calculate the forces and moments acting upon the breast – but now we want to determine the lengthening of the breast as well.

Guided by the elementary foundations of mechanics, we again postulate another brutal assumption: We will assume that the support reactions and the stresses calculated for the rigid body deviate only very slightly from those of the elastic body! This is not an obvious step. Consider the cantilever loaded by a weight force. The stress calculation has shown that no normal force acts on the cross section of the beam. When there are significant deformations – take an elastic ruler for verification of this – the end of the beam tilts in the direction of the weight force. The greater the angle of inclination means that a greater normal force is introduced into the beam! We will ignore this influential factor in the following.

2.1 The "Who is Who" of Material Strength: Stress, Strain and Modulus of Elasticity

Up until now the cross-sectional dimensioning of our members, for example the nipple, has not come into play. We will now put an end to that. And since there is not always a breast at hand for experimenting on your own and since this book has to pass the critical eye of the Juvenile Protection Board (and the even more critical eye of Dr. Hinrichs's mommy), instead of the breast we'll concentrate in the following on three bungee jumpers, who have not read the small print in the "Jet and Jump offer", where there is obviously no legal clause covering the subsequent rescue of the customer.

We would like to take another look at the respective stresses in terms of material strength. From statics we still know that normal force F_N = mg acts on the small isolated rope element. In statics, we left the cross-section of the members out of our calculations. However, there is no denying the fact that the cable exhibits spatial elongation in its cross-sectional area A – and that this cross-sectional area does indeed have a decisive influence on our pulse rate before the jump, i.e. the loading of the cable really does have something to do with the cross-sectional area!

The normal force F_N has to act somewhere at a point in this face – but where exactly? Well, everywhere of course, thus with N small forces having the magnitude F_N / N, yielding a total sum of F_N. So we can actually accommodate an infinite number of forces (N→∞) in the cross-sectional area (with small forces F_N / N = ...).

Thus the concept of forces is no longer adequate for describing the processes occurring in the cross-sectional area. The load exerted on the surface is apparently a function of surface area on the one hand, and of the applied force on the other. Here we will have to introduce a new unit of quantity:

$$\text{stress: } \sigma = \frac{F}{A} \qquad [N/mm^2]$$

Explanation:

This funny symbol at the beginning is not some new Pokémon figure, but rather the symbol for stress. It's pronounced "sigma", meaning that it is a Greek s!

It thus follows that the load exerted on the cord remains the same if the mass of the bungee jumper (or even better, dangler) is doubled for a cord having twice the cross-sectional area.

Since the small force-arrows in the cross-sectional area are all perpendicular, and thus "normal", to the surface area, this type of stress is also referred to as normal stress – this system of notation thus corresponds to the designation of the normal force, which is of course the resultant stress factor in our cord. Normal stress always has the effect of lengthening (or shortening) of a small element of the body, which in our example results in the lengthening of the cord.

The magnitude of the tension thus indicates the actual load acting on the member. The smallest particles or molecules of a jumbo jet and those of a needle, both subjected to the same stresses, will therefore be put under the strain of equal load forces. In this case, these very different members will even exhibit the same deformation – as well as the same risk of a malfunction....

Let us now turn to deformations, which apparently have something to do with stresses:

Back to our bungee jumpers, who are "hanging around together" (Dr. Romberg is good at hanging around even without a bungee cord!). The cord of the featherweight beauty (weight force G) is deformed to an appreciably lesser extent (i.e. lengthened by the amount x) than that belonging to the hefty regular of the iron-pumping shop.

Precise measurements made under laboratory conditions will then show that if two beauties (weight force 2 G) jump together on one cord, it will undergo a double lengthening 2x. A scientific evaluation of this gives the following resulting-diagram-chart-plot-evaluation:

In the diagram of tensile force F plotted with respect to lengthening x, we can see the range corresponding to this linear proportionality (double tensile force ==> double elongation).

The funny sign in front of the L at the end of the x axis is neither a tent nor a triangle, but a "delta". It describes the change in length, thus $\Delta L = L(N) - L_0$ (change in length = length at normal force N minus unloaded initial length).

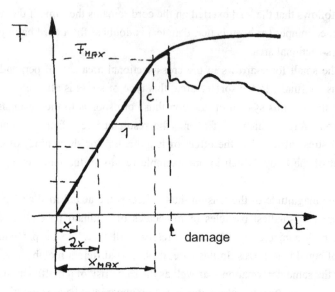

Figure 42: Tensile force F as a function of lengthening ΔL

This proportionality is valid until the cord snaps or somebody's skull smashes onto the asphalt – then we lose experimental control. We also lose control of our shrewd calculations, so we must therefore limit ourselves to the linear range. This is all normally summed up in the theory of springs, also called Hooke's Law:

$$F = c\, \Delta L.$$

Since the preceding has been pretty understandable to any house(man)wife with a high-school diploma, it's time to rev up the scientific fog machine: Experience shows that

- the doubled weight (F = 2G) results in the doubled hanging length, i.e. we deduce with razor-sharp acumen: ΔL ~ F (one reads: The change in length ΔL is proportional to (tensile) force F),

- if we double the length L of the rubber band, the lengthening is also doubled: ΔL ~ L,

76

– doubling the cross-section A of the rubber band (or two rubber bands) for the same weight results in only half of the lengthening, i.e. $\Delta L \sim 1/A$,

– the stiffer the material used in the rubber band, the less excursion is caused by the weight force. Here we refer to the stiffness of the material as ... here ... as ... well, let's just take ... E, so therefore $\Delta L \sim 1/E$.

So for the excursion ΔL (according to simple linear mass law) is can be shown that:

$$\Delta L = \frac{FL}{EA} \qquad \qquad \heartsuit$$

Instead of force, we now use stress as the reference value:

$$\sigma = \frac{F}{A} \qquad \qquad [N/mm^2] \quad .$$

Stress therefore indicates the force per surface element of the tension member or cord. Instead of the lengthening ΔL we will now make things even more foggy by introducing tennnnnnsion, or the lengthening per portion of the cord:

$$\varepsilon = \frac{\Delta L}{L} \qquad \qquad [1].$$

In the diagram we have given the spring constant c (in terms of a length) a really complicated name: the modulus of elasticity E (or, among friends, the E-modulus). Sounds good, huh?

The scientific fog machine has really picked up steam here: The technical subject matter has remained the same, as you can see from the identical illustrations – but the designations used now require a small dictionary for the inexperienced still-loser.

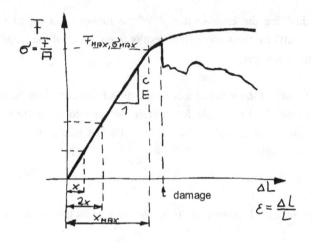

Figure 43: As in Fig. 42, but now with ultra-scientific axis notation

As a result of altering the units with the introduction of σ and ε, the unit of the proportionality constants has also changed: The elasticity modulus E is expressed by the unit N/mm². This is a material-specific constant, i.e. its value is given for each material upon purchase, regardless of color, chosen diameter..., and is thus simply a function of the material itself. Now you can quickly transform the spring theory: The previous equations can be expressed as

$$\sigma = \varepsilon\, E \qquad .$$

The model according to which deformation (tension ε) is dependent on loading (stress σ) is called mass law – in the case at hand we applied a linear mass law to a linear elastic range and that's the way it's going to stay!

A nesting of the equations reveals the following "vital" facts: The equivalent stiffness of an elastic body can also be calculated from its material and geometric data:

$$\Delta L = \frac{FL}{EA} \qquad ,$$

$$c_{\text{equi}} = \frac{EA}{L} \qquad .$$

78

To be honest we should note here that we have not only achieved the desired "fogginess" in the original spring theory but also a definite advantage:

According to the simple spring theory, a thin cord naturally has a different stiffness than a thick one. A long rubber band elongates more under the same weight than a short one. . . . With the help of the new equations you can dimension a member if you know its geometry and the elasticity module E as its material constant. Now that's something, isn't it?

2.2 Stress and Tension Under Normal Force and Simultaneous Warming

As everyone knows, mechanical engineers from the South possess somewhat more fiery temperaments than those from the cool North. So naturally this is something we have to take a critical look at: Under the high temperatures in the sweltering South, the same tensions between people have graver consequences.

In the meantime our bungee dangler is pretty well hung in the midday heat of the next day – he is no longer able to really notice the rise in temperature (his bloody hands are the result of his many desperate climbing efforts). But the cord definitely notices the change in temperature: It becomes longer when heated. So in our formula for the lengthening of the cord we will also have to account for the cord's "temperament" under the influence of increased temperature:

$$\Delta L = \frac{FL}{EA} + \alpha \, \Delta\vartheta \, L$$

where α: coefficient of thermal expansion. This is a
material-specific constant describing the
sensitivity (temperament) of the cord to a
rise in temperature,

$\Delta\vartheta$: the change in temperature.
Pretty obvious here: The more it heats up,
the longer it gets!

However, the development of a rather rank smell creates another problem for the bungee jumper – our calculations must now account for the vultures which are comfortably perched on the cable. The experienced statics engineer will immediately recognize that the normal force F_N is no longer constant but increases across the length (with increasing number of vultures). This means that we cannot use our formulas ♥ and ♠*, since they do not allow us to calculate with a normal force dependent on a specific location.

But since none of this is really all that complicated, we would like to generalize this example somewhat: In the following, our cord, in addition to the weight of the bungee-dangler, will also be loaded with its own weight, thus a distributed load. The entire apparatus is then warmed up – of course not uniformly in all places. That much is clear. Oh, yeah: We'll additionally assume that, for some inexplicable reason, the cross-sectional area and the coefficient of expansion of the rubber band are not constant along its length. This happens quite regularly in actual practice!

The magic formula is now:

$$\Delta L = \int_0^l \left(\frac{\sigma(x)}{E(x)} + \alpha(x)\,\Delta\vartheta(x) \right) dx \,, \qquad \text{☠}$$

where

$$\sigma(x) = \frac{N(x)}{A(x)}$$

designates the stress that we're now sufficiently acquainted with. Whoever refuses to believe this formula... will just have to believe it anyway, because we're always right, of course (after all, we are erudite scholars.).

So that's not bad at all, is it? But watch out: One thing you must always check is which assumption has been made in the formulation of the problem, otherwise you'll be stuck with much more work than necessary:

case 1: constant load, constant cross-section, *no warming*
==> equation ♥

case 2: load, cross-section, *warming* and expansion coefficient constant
==> equation 💣

case 3: ***everything is open***
==> equation ☠

Here's a tip from Dr. Hinrichs for the hot-shots: Of course you can always beat cases 1 and 2 using the more general case 3[14]...

So let's nail down these three cases right away with some examples:

Hanging downward L+ΔL by his tie (cross-sectional area A, E-modulus E, specific gravity[15] γ[16], coefficient of thermal expansion α) is a suicide (cadaver mass G, ~~flunked mechanics finals three times in a row~~).

a) The rope (noose) is massless (γ=0).
b) The own weight of the rope is to be accounted for.
c) The temperature of the rope increases from an ambient temperature of 40 degrees in spring to 70° in summer.

By what amount ΔL did the rope lengthen?

Given: G, γ, α, A, E, L.

---

[14] Dr. Romberg thinks that you would be just as well off beating yourself up!
[15] Or weight of unit volume.
[16] Here the editor would like to point out that according to the Industrial Standards, in effect for the last 25 years or so, the term unit weight has been replaced by specific gravity.

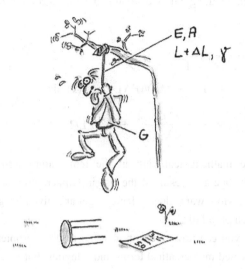

Nothing to it, right?! For problem part a) we immediately decide to take case 1, and thus equation ♥. But this provides us right away – without any great calculating work and by simple copying – with the solution:

$$\Delta L = \frac{GL}{EA} \quad .$$

b) We first have to deal once more here with statics, since here, only the normal force is needed for the rope. This is G at the lower end of the rope, while due to the rope's own weight the normal force at its upper end is G +γAL. Well, and what about in between? Here the normal force increases in linear fashion, since with every centimeter of rope, the weight of each additional centimeter exerts an additional pull on the rope, thus resulting in

$$N(x) = G+\gamma Ax.$$

Now we'll have to work our way through the integral equation ☘:

$$\Delta L = \int_0^L ((G+\gamma Ax)/EA) \, dx \quad = \frac{GL + \gamma AL^2 / 2}{EA} \quad .$$

c) Finally, the whole thing all over again in short form for the additional warming:

$$\Delta\vartheta(x) = 15° \frac{x}{L}$$

$$\Delta L = \int_0^L (G+\gamma Ax)/(EA) + \alpha\Delta\vartheta(x))\, dx$$

$$= \frac{GL + \gamma AL^2/2}{EA} + 15° \,\alpha\, L/2 \qquad .$$

Here the positive mathematical sign in the temperature term indicates a lengthening of the rope as a result of the rise in temperature! Bonus question from Dr. Romberg: How warm does it have to get at a given hanging height H for the suicide attempt to fail?

In this spirit you can now think up all sorts of great problems that yield somewhat complicated mathematical terms and integrals, but that really do not mean much in the way of providing new knowledge for the average loser: variable cross-sections (e.g. thick ropes tied to each other with a knot), wildly fluctuating temperatures with an additional fire under the ass) ....

2.3 Stressed Out in All Directions: the Circle of Stress

2.3.1 The Single-Axis Model: Rod under Normal Force Stress

Let's play "Battleship". You take a pencil with a pointed end and an eraser at the other end. If you place the pencil upright with its eraser on a sheet of paper and press on its tip (but be verrry careeeeful: It's pointed, man!), then a force F will act on its tip, or at a point.

While continuing to exert force we now tilt the pencil at an angle α to the vertical. Of course the first thing we have to do is to hold the sheet of paper under the pencil in place, otherwise the entire experimental apparatus will slip off sideways.

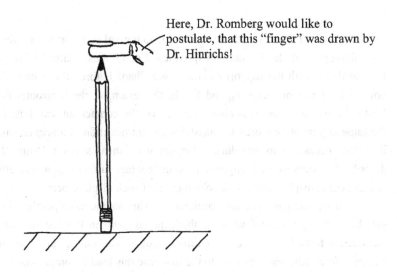

Here, Dr. Romberg would like to postulate, that this "finger" was drawn by Dr. Hinrichs!

Figure 44: Battleship: initial position

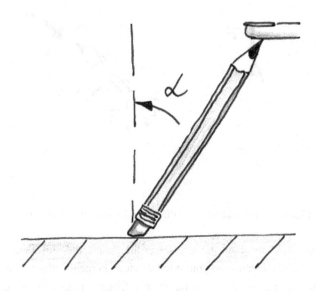

Figure 45: Battleship: final position

In addition to the shortening of the eraser due to normal stress, another effect is also taking place under the altered experimental conditions: a lateral migration. In accordance with the support reactions, see Chap. 1, force F can be broken down into the components F_H and F_V. In this example, the horizontal force (that's the frictional force) is also generated on the contact surface. Based on the same argumentation used for justifying the introduction of a normal stress, it is now necessary to introduce a tangential or shear stress τ [N/mm²] to describe the stress exerted tangentially to the contact surface and whose effect is a tangential displacement or the obstruction of such displacement.

Actually, the more you look at these experiments the more perplexed you get: In both experimental variants, the external load on the rod (pencil) is identical: A force F is applied in the direction of the rod's axis, thus generating a normal force with magnitude F. In the first case this load is compensated by a normal stress, in the second case by an appropriate combination of normal and shear stress. This actually functions analogously to the resolution of a force into a force couple, see Statics in Chapter 1.

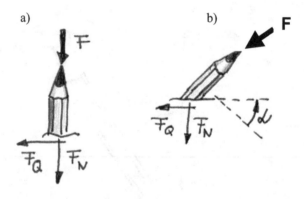

Figure 46: Internal forces of the pencil for a) α=0 and b) α≠0

So if anybody asks us about the stresses in the pencil, the only thing we can really say is: It just depends on your point of view. Perpendicular to the axis of the pencil we have a normal stress, but from a different point of view, or angle of cut, we have a combination of shear stress and normal stress, each of which is just large enough in magnitude to keep the external force in equilibrium. But this, of course, is somewhat vague.

As in statics, we would like to calculate the exact stresses for our pencil. But to do so, we make the cut of our section at two different angles: Case 1: as usual, thus perpendicular to the rod axis, and case 2: rotated from the normal cut by angle α[17].

Note: Surprisingly, the rod here, in contrast to the usual drawings encountered in statics, is turned upside down or rotated. This is not due to a mental blackout of the "draftsman" but is rather a deliberate analogy to our pencil experiment.

Having passed the chapter on statics with flying colors, the well-trained loser will of course immediately see that the stresses are found by

$$F_N \quad = -F \cos\alpha \,,$$

$$F_Q \quad = -F \sin\alpha \,.$$

Plotting the resulting stresses as a function of the cut angle α, you get. . . the following circle:

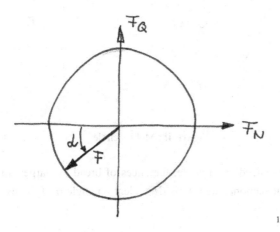

18

Figure 47: Normal force and shear force as a function of cut angle α

---

[17] What follows is an elegant mathematical deduction of Mohr's Circle, the author naturally being Dr. Hinrichs. Helpful tip from Dr. Romberg: Open up a nice cold beer, relax and resume reading tomorrow a few pages further on.

[18] This circle is a bit too oval for the editor!

Now we have to include the sectional area of the rod in our calculations, since it of course varies according to different cut angles. A complicated linear mathematical operation, dividing by the sectional area, turns the calculated force into a stress:

$$\sigma(\alpha) = F_N/A_S$$

and

$$\tau(\alpha) = F_Q/A_S \ .$$

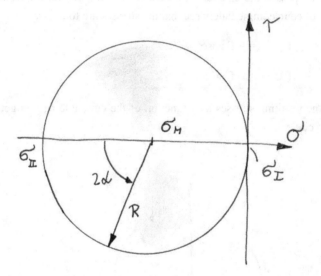

Figure 48: Mohr's circle

If a baguette is sliced at an angle, the slices of bread are larger. In the same manner, the cut sectional area A_S is dependent on angle α of the executed cut. It follows that if

$$A_S \quad = A/\cos\alpha$$

then

$$\sigma(\alpha) \quad = F_N/A_S$$

$$= F \cos^2\alpha \, / \, A$$

$$= \frac{F}{2A} \, (1 + \cos 2\alpha)$$

and

$$\tau(\alpha) \quad = F_Q/A_S$$

$$= F \sin\alpha \, \cos\alpha \, / \, A$$

$$= \frac{F}{2A} \, \sin 2\alpha \ .$$

This results in a circular-shaped figure, which we would like to glorify somewhat with the scientific term of Mohr's circle!

See how easy it can be to calculate your way through the horrors of materials strength? This has been done here for the pure normal force stress, the so-called single-axis stress .

In the case of our loaded pencil, the circle has the following characteristics:

1) center: $\sigma_M = \dfrac{F}{2A}$, $\tau_M = 0$

2) radius: $R = \dfrac{F}{2A}$

I like both the tensile and the shear stresses!

chrackckck

chruuunch

snap...creaaack

3)   maximum normal stress:
$$\sigma_{max} = -\frac{F}{A} \quad \text{for } \alpha = 0$$

(you probably could have guessed this result without any calculations, couldn't you?!)

4)   maximum shear stress: [19]
$$\tau_{max} = \frac{F}{2A} \quad \text{at } \alpha = 45°$$

So, pay close attention here: When a cut is made through a loaded body, the stresses of the sectional area are dependent on the angle of the cut. The normal stresses at those points of Mohr's circle where the shear stress τ disappears, i.e. the points of the circle that intersect with the σ axis, are also referred to as principal stresses (in our example with the pencil: $\sigma_I = 0$, $\sigma_{II} = -F/A$). Maximum stress occurred in the direction of the only loading factor. Perpendicular to the loading direction, the load is zero. But this has been pretty clear to us all along: For $\alpha = 90°$ we determine the tension parallel to the side surface of the rod, thus slicing open the pencil in its longitudinal direction... Since these surfaces are not longitudinally loaded, there is no reason for any stresses to be present here.

The case we have just examined involves what is referred to as single-axis stress. In the case of two- (three-) axis stresses, there are two (three) loads having components in two (three) directions.

Properties of Mohr's circle (Part I)
1) The centers of any stress circle lie on the σ axis!
2) The angle α at the member is plotted in the stress circle under the angle 2α
3) Determining positive and negative signs:

---

[19] (Here a small tip for the experienced practician or materials scientist: It is for this reason that tensile test samples of ductile = free-flowing materials tear at an angle of 45° to the rod axis when the maximum shear stress is exceeded. When brittle materials exceed the maximum normal stress, their fracture surface lies perpendicular to the rod axis.)

a) Positive normal stresses correspond to tensile stresses, negative normal stresses correspond to compressive stresses.

b) Shear stress is plotted in Mohr's circle as follows: The normal stress σ points out of the sectional area... If you have to rotate the latter clockwise in order to make it coincide with the shear stress τ, then τ is positive. Otherwise τ is negative.

4) The principal stresses are sorted according to magnitude:

$$\sigma_I > \sigma_{II} \;(> \sigma_{III}, \text{ three axial case})$$

5) The theorem of associated shear stresses:

"The shear stresses in two perpendicular sectional areas have the same value."

For the experienced Mohr's circle freak, this is a trivial point, since these points must of course lie opposite one another in the circle $(2 * 90° = 180°)$.

Experience has shown that the greatest, but most easily avoided, source of error is actually point 3) concerning positive or negative signs.

If we know that the stresses involved can be plotted in the form of Mohr's circle, there is no longer any need to use the previous equations at all. To construct Mohr's circle, it is quite sufficient to know just two points of the circle, for instance, knowing:

a)   the principal stresses $\sigma_{II}=0$ and σ_I where $\tau = 0$ (for single-axis stress),

or

b)   that for any partial area of the member, the normal and shear stress acting on this area are known for single-axis stress.

Here's a quick example:

Dr. Romberg has to have a tooth (cross-sectional area A) pulled. As a result of the tensile forces acting on the tooth (and also due to his inflamed smoker's gums) Dr. Romberg is in great pain. The shear stress τ and normal stress σ act on the root of the tooth at an angle α, as shown in Figure 49[20]. (The forces (stresses) in the lateral areas of the tooth can be disregarded, according to information provided by the dentist.)

a) After assessing the pain-inducing stresses τ and σ, Dr. Romberg tries unsuccessfully to assess the maximum shear stress acting in the tooth. Can Dr. Romberg be helped?[21]
<u>Solution:</u>   No.

b) With what tensile force F does the dentist pull on the tooth?[22]
Given: σ = 200 N/mm², τ = 100 N/mm², A = 20 mm².

Figure 49: Tooth under angle α

---

[20] Another thing Dr. Romberg still can't get out of his head is the idea that the Americans never landed on the moon but instead staged the landing on a Hollywood film set. We'll probably never be able to escape it – during discussions on this topic he loves to provide *detailed* contradictory evidence, for example that the light reflecting off the studio space ship was set at the wrong angle

[21] Here we're only asking about help in the CALCULATIONS – any other kind of help would be too late anyway!

[22] Altered formulation of the problem: How hard does Dr. Hinrichs have to pull on the tooth in the given example for Dr. Romberg to experience a maximum pain load at the impending tooth extraction?

With the known points P_1 and P_2 we can now fiddle around with the Mohr's circle...

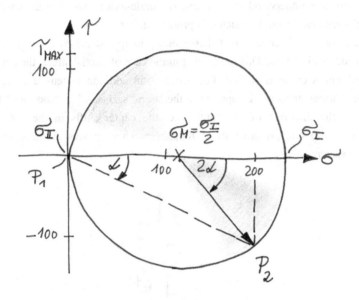

Figure 50: Mohr's circle (well, anyway)

The maximum shear stress and the principal stress practically jump out at us:

σ_I = 250.000000000043175 N/mm²,

τ_{max} = 125.0000000000000002143 N/mm².

Since it would take a more sophisticated drawing for us to read any more decimal places, we will supply the calculated solution here:

α = arctan (100 N/mm² / 200 N/mm²) ,

τ_{max} = R = 100 N/mm² / sin(2α) ,

σ_I = 2 R .

The tensile force on the tooth is then found by

$$F = \sigma_I A = 5.0 \text{ kN} \quad .$$

2.3.2 Biaxial Stress

Once you have understood the principle of single-axis stress, then the case of biaxial stress should not be much of a problem either.

The only difference is that the second principal stress σ_I (or σ_{II}) is no longer identical to zero. This means that in the case of biaxial stress, the center of the Mohr's circle is shifted. For instance, in addition to tensile force F, another normal stress can be applied to the lateral surface, thus "zooming" the circle in the direction of the "pull", i.e. the center shifts in the positive direction, with one principal stress remaining where it was in the case of single-axis stress.

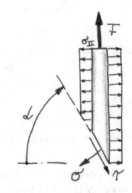

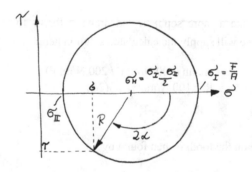

Figure 51: a) FBD of a "four wheeler" (i.e. two axles)   b) Mohr's circle

Simple, right? Unfortunately, for this type of stress one can think up problems that are much more devious than this, see Chapter 4.2.

2.3.3 The triaxial model

As we have seen, the single-axle model is a special case of the biaxial one. Exactly the same applies to triaxial stress, which should be old hat by now. Here you can always observe the actual conditions on an arbitrary surface of a body by analyzing a cuboidal element of it – but as in the case of biaxial stress these can be represented as a circle.

In order to examine all three spatial dimensions, we have to look at three faces of the cube – that is, three stress circles. So the resulting figure then looks like the following:

Understandably, the circles abut one another, giving us the three principal stresses σ_I, σ_{II} and σ_{III}. That's all we need.

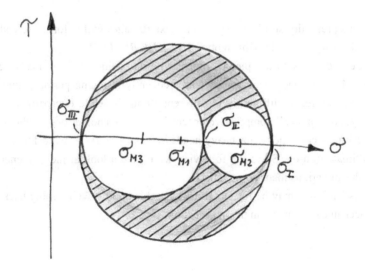

Figure 52: Triaxial stress

As a special treat – in his opinion – Dr. Hinrichs has even drawn up a nice little table!

Properties of the Mohr's circle (Part II)			
		Type of stress	
		single axle	biaxial	triaxial
Principal stress σ_I	σ_I or $\sigma_{II} \neq 0$, e.g. F/A	$\neq 0$	$\neq 0$
σ_{II}		$\neq 0$	$\neq 0$
σ_{III}			$\neq 0$
Radius R	$\sigma_I/2$ or $\sigma_{II}/2$	$(\sigma_I-\sigma_{II})/2$	$R_1=(\sigma_I-\sigma_{III})/2$ $R_2=(\sigma_I-\sigma_{II})/2$ $R_3=(\sigma_{II}-\sigma_{III})/2$
Center σ_M	$R=\sigma_I/2$	$\sigma_{II}+R$	$\sigma_{M1} = \sigma_{III}+R_1$ $\sigma_{M2} = \sigma_{II}+R_2$ $\sigma_{M3} = \sigma_I+R_3$

After this pretty dry stuff – *sorry !* – the next 10 pages will unfortunately also stay just as dry. So stick to this motto: *BITE THE BULLET*!

Let's go back to our pierced nipple with its additional S&M load. Here the S&M practitioner, experienced with frequently changing partners, knows that everyone reacts differently to different stimuli – while one partner may really get off on small compressive or tensile forces – another one might need tensile forces combined with heavy-handed caresses with the whip for his or her ultimate turn-on. So in the following we'll take a look at the influencing variables of various loads on members.

As hard as it may be for Dr. Hinrichs at this point – we're going back to our mechanical members in the next section.

2.4 Comparative Stresses

Some structural members exhibit highly sensitive reactions to a normal stress, i.e. they fail when a critical normal stress is exceeded. Others reach the climax of inner tear when a critical shear stress is surpassed. And again others take a more middle-of-the-road approach. So for the design of these members there naturally have to be different criteria that reflect these sensibilities and preferences of the material involved.

A material failure (a tear or crack) of the structural member caused by exceeding the admissible normal stress for so-called brittle materials is referred to as a brittle fracture, which always runs perpendicular to the load direction, thus perpendicular to the maximum normal stress.

The most elementary evaluative model would involve the following procedure:

1)   Since the magnitude of the maximum occurring normal force is always at least as large as the maximum shear stress, we simply compare the maximum normal stress with the maximum allowed stress specified for the material (and hope that the shear stress does not upset our calculations). This is what a scientist always means when he refers to a normal stress theory.

2)   For a second group of materials (try a piece of chewing gum some time!) the fracture occurs along a surface extending at a 45° angle to the direction of the maximum normal stress – these are ductile materials that fail when the maximum permissible shear stress is exceeded. (Maximum shear stress in Mohr's circle rotated by 90° to the principal stress, by 45° as a fracture.)

And between the two there are all sorts of in-between kinds of things: those which are ductile to a greater or lesser degree, etc. We therefore have to think of some sort of criterion for evaluating the occurring stresses. And then apply that in determining a "comparative stress" σ_V, put together from the shear and normal stresses, so we can then compare it with the admissible stress for the chosen material. But since this seems pretty complicated even to us, we will just copy things down here and believe the not-so-new theories:

3) Trescasche's flow criterion: Here we try to find the maximum difference between an arbitrary combination of principal stresses, thus

$$\sigma_V = \max(|\sigma_{II} - \sigma_I|, |\sigma_{III} - \sigma_{II}|, |\sigma_{III} - \sigma_I|)$$

4)   The deformation theory (sounds so good we'll even add something to it: also called the HUBER-MISES-HENKY flow criterion)

$$\sigma_V = \sqrt{0.5[(\sigma_I - \sigma_{II})^2 + (\sigma_{III} - \sigma_{II})^2 + (\sigma_{III} - \sigma_I)^2]}$$

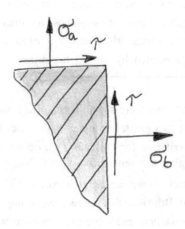

Figure 53: Stresses at perpendicular faces

The disadvantage of these formulations is that the principal stresses must be provided or calculated. With a lot of trigonometry you can also determine these comparative stresses at two surfaces rotated by 90° – the mechanics are the same!

2a) Normal stress hypothesis

$$\sigma_V = 0.5 \, |\sigma_a + \sigma_b| + 0.5 \, \sqrt{(\sigma_a - \sigma_b)^2 + 4\tau^2}$$

3a) Shear stress hypothesis

$$\sigma_V = \sqrt{(\sigma_a - \sigma_b)^2 + 4\tau^2}$$

4a) Deformation hypothesis

$$\sigma_V = \sqrt{\sigma_a{}^2 + \sigma_b{}^2 - \sigma_a\sigma_b + 3\tau^2}$$

This manner of representation makes it even more obvious that the occurring normal and shear stresses in each hypothesis are evaluated differently. On the other hand, of course, these are all model calculations which hopefully have a lot in common with reality.

We trust that you are convinced by now that deformations occur in a tension rod as a result of the normal forces and stresses occurring in the rod and that these deformations are manifested in the lengthening or shortening of the rod. As you already know, in addition to normal forces, shear forces and bending moments can occur as potential stress factors in a member. In the following we will take a closer look at the influence of these load factors on the resulting deformations.

2.5 Bending of Beams

We would like to study the limiting quantities of geometrical and material parameters on deformation in a simple experiment.

2.5.1 The Geometrical Moment of Inertia

Taking an elastic ruler, you hang a weight (e.g. the coffee cup) from one of its ends by a string. At the other end of the ruler, your hand will simulate a built-in support by bringing the ruler into a horizontal position (with your hand introducing a force and a bending moment into the support). If you then apply the weight force on this cantilever in doses while hoping that the bending stress in the support does not exceed the maximum admissible stress, then you can conduct the following series of experiments:

Cases of loading, when:

1) the ruler is held like a "diving board", i.e. with horizontal orientation of the broader side,
2) the ruler is rotated 90° about its longitudinal axis,
3) a position at an arbitrary angle between the two extreme positions 1) and 2) is chosen!

It can be seen that the bending in load case 1) is much greater than that in load case 2), although of course we didn't change anything concerning the material data or geometry of our support. While the practitioner will not be surprised by the results of tests 1) and 2), he will be puzzled by test 3) where the support not only bends in the direction of the weight force but is also bent horizontally, i.e. it migrates sideways. Of course this does not surprise an experienced mechanical engineer (after making a few calculations) – this is old hat: skewed bending. But since we are all novice mechanical engineers, let's make the fictive assumption that for now skewed bending does not occur in practice.[23] Elongation under pure tension (see Section 2.1) is of course proportional to the cross-sectional area of the support. But the bending of the support does not appear to be directly dependent on its area. But then which influential quantity is responsible for greater deflection in case 2?

Here's another experiment: If you try to lay a piece of paper between the edges of two tables, the paper will sag quite considerably. Even successful

---

[23] For all hot shots: Refer to the textbooks and/or study the problem on skewed bending in Chapter 4.

learning processes during early childhood development suggest that this paper support will stay in place better if it is fan-folded a few times.

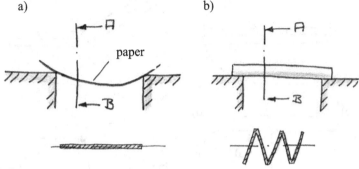

Schnitt A-B:

Figure 54: Deflection of a a) non-folded and
b) folded sheet of paper[24]

Now for the prize question: What is better about the folded sheet of paper than the smooth one? If you take a look at the cross-sections of the two paper supports, you will notice the main difference is that the material in the case of the folded paper has been brought out further away from the drawn middle line. But why does it seem to be better when the material is arranged further out? Here is another small sketch showing a model of a small element of our paper:

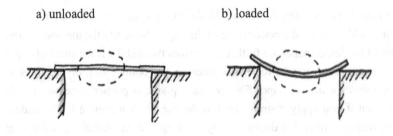

Figure 55: Paper: a) unloaded   and b) loaded, e.g. by its own weight

---

[24] Here we have purposely neglected any compacting of the paper resulting from the folds.

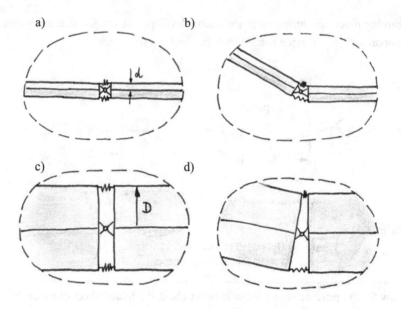

Figure 56: Model of a small paper element:
a) unloaded, not folded,
b) loaded, not folded,
c) unloaded, folded
d) loaded, folded

The three horizontal lines of the model designate a paper fiber on the top side, in the middle and on the bottom side of the paper. Naturally the top and bottom sides of the folded paper are further away from the middle. If the sheet of paper is then bent downwards, the paper fiber on the bottom side of the sheet is pulled apart, while the paper fiber on the top side is pressed together (this is best seen if you apply a strong load to the paper). Somewhere in the middle there will be a fiber that doesn't change in length. In the following we'll refer to this fiber as... the neutral fiber, or something like that.

The respective lengthening and shortening of fibers are necessary in order to compensate for the bending moment, which of course acts as the internal force in the support. But here the folded support has two advantages:

1) Due to the greater distance d in the folded support, the same twisting of the support's cross sections results in a greater tensile force (compressive force) in the spring at the bottom side (top side) of the support, F ~ d.

2) The lever arm d of this generated reaction (tension / compression) to bending is likewise greater in the folded support, $\Rightarrow M_{BACK} \sim F\,d \sim d^2$.

So to put everything in a nutshell:
Bending seems to depend on the square of the distance d of the sectional area elements A from the middle line (the neutral fiber). The quantity we would like to model this effect is referred to as the geometrical moment of inertia I with I ~ A d² [mm⁴][25].

In a somewhat more realistic support, such as a beam, the individual springs shown in Figure 56 blur into an infinite number of small springs. The stress distribution resulting from bending then looks something like this:

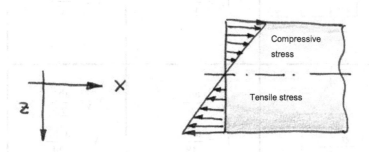

Figure 57: Stress distribution across the beam's cross section

Also of importance in some applications is the so-called "section modulus W", which makes it possible to calculate the maximum stress "furthest out". This is defined as follows:

$$W(x) = \frac{I(x)}{|z_{max}|} \qquad ,$$

---

[25] (The word inertia comes from the Latin *iners*, or "inert", an apt description of the brain activity of students when first introduced to this concept.)

where $I(x)$ is the geometrical moment of inertia and $|z|_{max}$ the point of maximum tensile stress furthest out on the loaded member (where the first fracture occurs). $|z_{max}|$ is usually the radius of a circular cross-section or half the width or height of some body... but you can get by without it.

We can determine the geometrical moment of inertia I and thus the section modulus W with more or less complicated integral calculation – but why reinvent the moment of inertia of the wheel. The easiest way is to look it up in the appropriate table, e.g.:

Cross section	Geometrical moment of inertia	Section modulus
	$$I_{yy} = \frac{bh^3}{12},$$ $$I_{zz} = \frac{b^3 h}{12}$$	$$W_y = \frac{bh^2}{6}$$ $$W_z = \frac{b^2 h}{6}$$
	$$I_{yy} = \frac{ah^3}{36},$$ $$I_{zz} = \frac{a^3 h}{48}$$	$$W_y = \frac{ah^2}{24}$$ $$W_z = \frac{a^2 h}{24}$$
	$$I_{yy} = I_{zz} = \frac{\pi}{4}\,(R^4 - r^4)$$ small wall thickness δ: $$I_{yy} = I_{zz} = \frac{\pi}{8} d_m^3\, \delta$$ (full circle: $r = 0$)	$W_y = W_z =$ $$\frac{\pi}{4}\left(\frac{R^4 - r^4}{R}\right)$$ small wall thickness δ: $$W_y = W_z = \frac{\pi}{4} d_m^2\, \delta$$ (full circle: $r = 0$)

And now for a little quiz to test your progress. Please close any other books, place your schoolbag under your desk and remove the batteries from your calculator...

The test question is:

Drawn in the following sketch (by Dr. Hinrichs) are cross-sections of a number of supports, where the rectangles are supposed to be equal in area. Please sort – while completely ignoring the "J's" (?) in the "drawings" – the geometrical moments of inertia for bending about the sketched axis X-X from greatest to lowest.

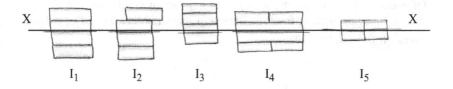

Figure 58: Different? geometrical moments of inertia I (!)

Solution: The rule of thumb is as follows: The greater the areas and the further these elements are located from the X-X axis, the less bending will occur (after all, the springs in Fig. 56 are located further from the X-X axis), thus giving a greater geometrical moment of inertia. This means that:

$$I_4 > I_3 > I_1 \ (I_1 = I_2) > I_5 \ .$$

This rule of thumb is also practiced in steel constructions, where supports are naturally not made using thin sheets of rolled steel, but instead with so-called I-beams, for example, where as much steel as possible can be arranged further out:

Of course, we can obtain the geometrical moment of inertia for such a body from the manufacturer or from the Bureau of Industrial Standards: $I_{XX} = 1140$ N/cm^4 (IPBv 100, DIN 1025).

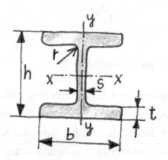

Figure 59: I-beam

But what do we do if we've just lost this information or if we happen to be in Egypt – with only the book *Don't Panic – It's Only Mechanics* in our bags – and want to build a pyramid using I-beams? In this case we could grossly simplify and use rectangles to compose the I-beam, which we could then find in our table for geometrical moments of inertia.

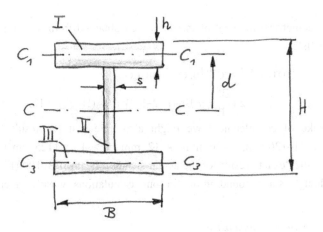

Figure 60: Equivalent model

For bending of the upper rectangle I around axis C_1-C_1 we apply:

$$I_{I, C_1C_1} = \frac{Bh^3}{12}.$$

But oh no! Now we're not going to bend the upper rectangle around its middle axis C_1-C_1, but instead around axis C-C running through the centroid of the beam. Here we can refer to the Huygens-Steiner theorem:

$$I_{CC} = I_{C_1C_1} + A\,d^2.$$

So when the neutral fiber is displaced by the distance d, the geometrical moment of inertia increases in accordance with the term $A\,d^2$, familiar to us from our exercise with the folded sheet of paper.

The bending of this partial body I about the C-C axis running through the centroid is expressed by the Steiner theorem:

$$I_{I, CC} = I_{I, C_1C_1} + A\,d^2 = \frac{Bh^3}{12} + Bh\,(H/2 - h/2)^2.$$

For rectangle II we can go right to the table for its value:

$$I_{II, CC} = s\,(H - 2h)^3 / 12 \qquad .$$

The total geometrical moment of inertia is then obtained from the sum of the individual "Is":

$$I_{TOTAL} = I_{I, CC} + I_{II, CC} + I_{III, CC}$$
$$= 2 \left(\frac{Bh^3}{12} + Bh \, (H/2 - h/2)^2 \right) + s \, (H - 2h)^3 / 12.$$

For the sake of completeness we might also note that the result (where B=106 mm, H=120 mm, h=20 mm, s=12 mm $\Longrightarrow$ I = 1125 cm^4) fairly approaches the "exact" result for the actual standard beam IPBv100 (I = 1140 cm^4), although some rounding-off in our calculations wasn't taken into account.

And now something to memorize:

When employing the so-called "Steiner's share", always make sure that the bending axis is always shifted from the mid-point (for homogenous bodies: center of mass). The geometrical moment of inertia is minimal for the bending axis through the mid-point!
So if you want to make an arbitrary shift of a reference bending axis, you first have to make a shift toward the center of mass (negative Steiner's share) and then a shift from that point (positive Steiner's share), otherwise the result is complete applesauce!

Since there's not much that can shock us well-seasoned Steiner practitioners, it's time for an advanced crash course:

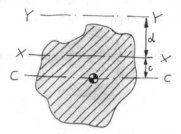

Figure 61: Sketch illustrating Steiner's theorem

Assuming that we know the geometrical moment of inertia of a body with respect to the X-X axis, we now want to find the same with respect to the Y-Y axis, see Figure 61.

Dr. Hinrichs still has not gotten over an incident that happened during his days as a young college seminar instructor when he presented the following equation, practically dripping with errors – he still frequently wakes up at night screaming, with terrible and unintelligible curses on his lips. But he still refuses to undergo any corrective therapy.

The equation he presented was:

$$I_{YY} = I_{XX} + A\,d^2 .$$

Well, if you haven't detected the mistake, there is one little thing you have not quite understood yet: As mentioned a few sentences above in italics, Steiner's theorem only applies when measuring from an axis running through the center of gravity (and back to this axis as well). The correct calculation should therefore be:

$$I_{YY} = I_{XX} - A\,c^2 + A\,(c + d)^2 \qquad .$$

And that is unfortunately something quite different that the result of the first clumsy attempt at finding a solution – otherwise the binomial theorem would have to be reformulated[26]. So remember: With Steiner *always* calculate away from and back to the centroid!

---

[26] Only Dr. Romberg would be able to keep working with his old version of the binomial theorem.

So that pretty much takes care of the geometrical moment of inertia, which in the following – and of course in practice as well – will always turn up in the given values.

But now we would like to go back to our original formulation of the problem: Which factors influence the bending of a support?

2.5.2 Bending

Leonardo da Vinci (1452–1519) already investigated the most important influencing parameters:

a) b) c)

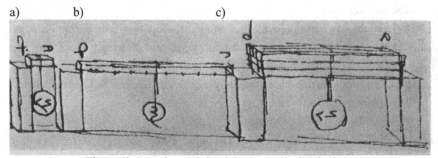

Figure 62: Experimental sketch by Leonardo da Vinci

Signore da Vinci not only experimented with various geometrical moments of inertia (see Fig. 62 b) and c)), but also with the length of the loaded supports. Here it should be clear to anyone that a long beam (Fig. 62b)) with a load is subjected to greater bending than a short beam (Fig. 62a)). Beam bending w is therefore a function of beam length L: $w = f(L)$.

Besides length L and the geometrical moment of inertia J, we can also nail down the modulus of elasticity, familiar to us from Chapter 2.1, as another parameter influencing the bending of a support. In our beam model from Fig. 56, E can be compared with the stiffness of the springs. Mechanics also refer to the bending function from the x coordinate of the beam function as the elastic line. So where do we get this elastic line? We take one of those nice tables for load types and look up everything in the table which looks "interesting":

Load cases	Elastic line equation	Bending	Inclination at beam end
	$w(x) =$ $\dfrac{F}{6EI} Lx^2\left(3 - \dfrac{x}{L}\right)$	$w(L) = \dfrac{FL^3}{3EI}$	$\tan\alpha = \dfrac{FL^2}{2EI}$
	$w(x<L/2) =$ $\dfrac{FL^3}{16EI}\left(\dfrac{x}{L} - \dfrac{4x^3}{3L^3}\right)$	$w(L/2)$ $= \dfrac{FL^3}{48EI}$	$\tan\alpha = \dfrac{FL^2}{16EI}$

That's pretty practical, isn't it?

As you can see, all of the results are dependent only on L, EI and, of course, load F or distributed load q. And for other load cases all you have to do is page through those great books listed in the bibliography.With the help of the table, we now can actually model every possible load case... Despite the apparently simple procedure of copying down the formulas, quite tricky problem exercises can be created from them, see Chap. 4.2. Unfortunately, not all load cases can be found in textbook tables! Here one might strongly suspect that such exceptions are much more frequent in exam questions than in actual practice, where as an engineer you can improvise a lot more in your calculations.

In the following we would like to explain how to derive the elastic line from arbitrary stress gradients.

For those of you who have had enough at this point, you are welcome to join Dr. Romberg in taking the shortcut to Chapter 2.6, marked by a

 !

2.5.3 Integration of the Elastic Line

If we go back to the model in Fig. 56 then it should be quite clear to us from statics that a bending moment is acting on the right and left side of each of these small elements – the internal forces (and moments). If we now take this imaginary little chain in hand, we can not only shift the chain in its horizontal position from top to bottom (displacement w), but also tilt it as a whole at an angle to the horizontal (angle w′).

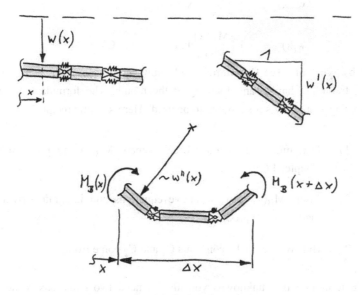

Figure 63: Models for bending of a beam

It is also possible for us to bend the chain by twisting our hands in opposite directions holding each end of the chain – we must introduce a bending moment into the support. The magnitude of this bending moment seems to be proportional to the "twisting" of the support. But the radius of the emerging curved shape is described by w″ (the second derivative based on point x). Thus for the entire support we get

$$w''(x) \sim M_B(x) \qquad .$$

Since the dependent variables derived for the folded paper support must still remain valid, the radius of the emerging curved shape will probably decrease with increasing modulus of elasticity E and increasing geometrical moment of inertia I. The complete formula is thus:

$$w''(x) = -\frac{M_B(x)}{EI(x)} \qquad .$$

So in order to determine the elastic line $w(x)$ of a support we have to doubly integrate the known right side of the equation and obtain the desired $w(x)$, thus

$$w'(x) = -\int \frac{M_B(x)}{EI(x)} dx + C_1,$$

$$w(x) = -\int\int \frac{M_B(x)}{EI(x)} dx\, dx + C_1 x + C_2 \qquad .$$

Too bad you can't see Dr. Hinrichs right now: slightly feverish with a lustful look, trembling hands, and foaming at the mouth. The formula looks very promising – but how's it supposed to be used? Here's a little recipe:

1) Determine the internal force moment $M_B(x)$ (no problem after Chapter 1)

2) Insert $M_B(x)$ into the above equations and integrate (trivial (?) mathematical problem)

3) But: Where do the constants C_1 and C_2 come from?

For determining two unknowns you always need two equations, which we obtain from the *boundary conditions*. There are geometric (position and slope, depending on the given geometry) and dynamic conditions (curvature, change in curvature). The latter depend on the applied loads.

Here a few examples:

Type of boundary condition	geometric boundary condition		dynamic boundary condition	
	w	w′	w″~M	w‴~Q
pinned and roller support	0	≠0	0	≠0
built-in support	0	0	≠0	≠0
free beam end	≠0	≠0	0	0
displaceable support	≠0	0	≠0	0

Got that? You should try to visualize these values. For example, for a built-in support: Assuming that the mason has walled in our support correctly, then it – regardless of its load – should project from the wall horizontally (w′=0). The wall under it should also not give (w=0). The support is held in place by the shear force provided by the wall $(Q \sim w''' \neq 0)$ and the bending moment $(M \sim w'' \neq 0)$.

Now we will try to run through the whole procedure to make sure that those old hands have provided the correct elastic line in Table 3.

The sketched plank (cantilever with length L, flexural strength EI) is loaded on its free end by force F (the weight of a captain going into a very brief retirement). Determine the elastic line of the support without looking at the table "Load cases and their elastic lines"!

Given: F, L, EI.

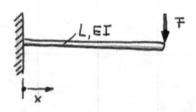

Figure 64: Cantilever

First the internal force moments: It follows that

$$M_B(x) = F(x - L) \quad .$$

The equation for the elastic line is provided by

116

$$w'(x) \;=\; \int \frac{F(L-x)}{EI}\,dx + C_1,$$

$$=\; \frac{1}{EI}\, F(Lx - x^2/2) + C_1.$$

From the table of boundary conditions we take

$$w'(0) = 0,$$

thus $C_1 = 0$. The next integration gives

$$w(x) \;=\; \int w'(x)\,dx + C_2$$

$$=\; \frac{1}{EI}\, F(Lx^2/2 - x^3/6) + C_2.$$

From the table (w(0) =0) we take $C_2 = 0$ and we can specify the elastic line for load case 1:

$$w(x) \;=\; \frac{1}{EI}\, F(Lx^2/2 - x^3/6) \;\;,\; w(L) = \frac{FL^3}{3EI}$$

We can therefore confirm the result in the table "Load cases and their elastic lines". The method we have just used can now be employed for all kinds of courses of moments and boundary conditions. This will sometimes result in terms longer than those in the example, and integration may have to be performed for individual regions (in case stresses have been introduced somewhere), or the integration constants do not happen to cancel out arbitrarily, see sample problems in Chapter 4.2. But the actual procedure is always the same as shown in the cantilever example above.

For the sake of completeness it should be pointed out here that of course you can always carry out the integration for a shear force Q or distributed load q:

$$w'''(x) = - Q(x)/(EI) \quad ,$$
$$w''''(x) = \; q(x)/(EI) \quad .$$

However, it makes sense to us to choose a consistent method for the solution when working with known stress gradients. Here we recommend solving all such problems by starting at $w''(x) = -M_B(x)/(EI)$.

So now we are able to calculate magnificent, aesthetically pleasing elastic lines. But let's go back once more to our stresses here. Stresses arise in structural members not only due to tensile forces or heat expansion (Chap. 2.1 f.), but also in conjunction with bending.

2.5.4 Stress Due to Bending

Let's now go back to our simple model of a small beam section: The springs above the non-stretched neutral fiber are pressed together (compressive stress) while those below are pulled apart (tensile stress).

Since the stress for the examples considered here is linearly dependent on the z coordinate, we can also describe its diffusion with the linear equation

$$\sigma(x) = \frac{M_B(x)}{I} z$$

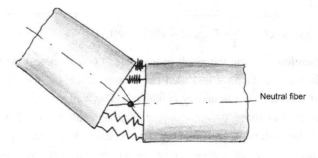

Neutral fiber

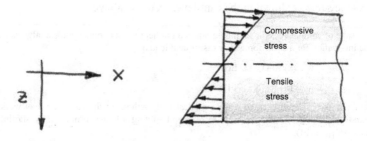

X

z

Compressive stress

Tensile stress

Figure 65: Stress distribution

118

in the bending beam. If an external normal force is now simultaneously superimposed on the bending stress, we simply add up the stresses:

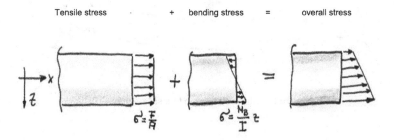

Figure 66: Superposition of a normal force load by a bending stress

The appropriate formula is thus:

$$\sigma = \frac{F}{A} + \frac{M_B(x)}{I} \, z \qquad .$$

"Any questions, Romberg?" "Yes, Hinrichs! When will we finally get through all this crap?"

2.5.5 Shear Stress Due to a Shear Force

It is just incredible when you think about everything that takes place during bending. We will consider the next effect for two slightly different support structures[27].

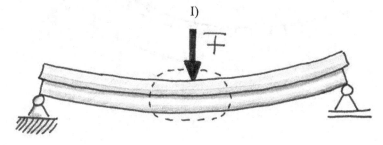

---

[27] Comment by Dr. Romberg: "No idiot's going to be interested in that..."

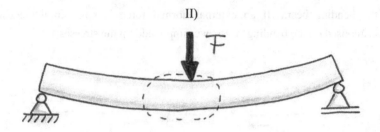

Figure 67: Shear stress due to a shear force

In example I two flat bars are laid on top of each other and loaded by force F. Because of the bending the end faces of the flat beams will of course no longer lie flush to one another. The situation is different in example II, where the two flat bars are glued to one another. Since the bars are glued together, their ends remain flush despite the induced deformation. The same effect can be achieved by increasing the friction between the two bodies. We would now like to put on a little peepshow of the friction contact between the two pliable, supple bodies: Oooooaaaaahhhhhhh!

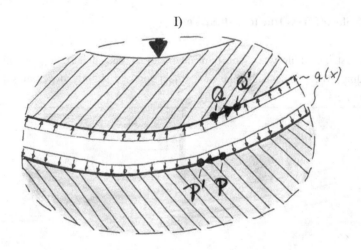

II)

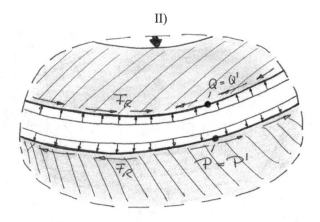

Figure 68: Peepshow of elastic, pliable bodies

Acting in the area of contact in example I is only a more or less constant distributed load q(x). The distance between the two contact points P and Q of the non-deformed support increases during deformation. The bottom side of the upper support is of course elongated due to the bending stress (Q ==> Q'), while the upper side of the other support is shortened due to bending stress (P ==> P'). But in the second example, the glue or the friction keeps the two points together even during deformation. This happens on account of the strength of the glue. (In our experiments we employed Elmer's, a dependable, high-performance glue – we don't know if other glues are just as suitable). Since the strength of the glue acts as a large frictional force, we have designated it as F_R. This glue force thus counteracts the deformation induced by bending. In the contact surface we must now refer to this result of bending as a shear stress, since the frictional forces act in the direction of the surface. But what happens in a support with the same dimensions as support II, but made of a single piece? Naturally the same thing happens. So for the record:

Bending stress entails simultaneous shear stress. And we can calculate this with the following formula

$$\tau(x,z) = \frac{Q(x)\,S(z)}{I\,b(z)} \quad .$$

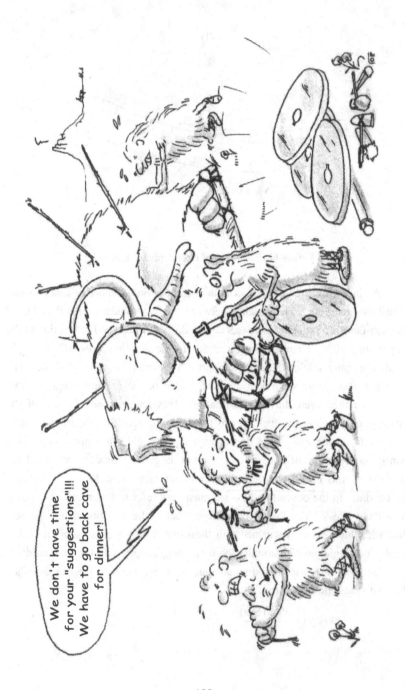

Again, you will just have to take our word on this formula if you want to keep up with the times!

Shear stress is therefore dependent on the shear force $Q(x)$ (and thus on the derivative of the bending moment $M_B(x)$). Since the shear force reverses the sign at the point where the force is introduced, the direction of the frictional forces F_R marked in the sketch also changes at this point. Furthermore, the shear stress depends on those ominous parameters $S(z)$ and $b(z)$, which in turn are dependent on the distance z from the neutral fiber. Here it's quite easy to determine the value $b(z)$: It designates the width of the support as a function of the z coordinate. But for $S(z)$ we again have to reach deep into our bag of tricks: $S(z)$ designates the moment of force. Well that's just great! But what is it?

Here we direct our attention to the profile in Figure 69. We want to determine $S(z)$ for the cross-section at point z_0. Now we only need the area A_{rest} of the cross-section lying below z_0 and the centroid coordinate $z_{S,rest}$ of this area. This gives us $S(z) = z_{S,rest} A_{rest}$.

Cross-sectional area

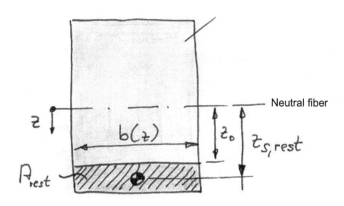

Figure 69: Moment of force

123

For the rectangular profile (B*H) taking

$$z_{S,rest} = z+(H/2-z)/2$$

and $A_{rest} = B(H/2-z)$

thus $S(z) = B[z(H/2-z)+(z^2 -Hz+H^2/4)/2]$.

Note: As can be seen in the example of the two glued beams, the calculated shear tension acts along the glue surface in the longitudinal direction of the beams. As a result of the theorem of associated shear stresses the same shear stress also acts in the z direction in the beam cross-section.

And now something for all hot shots. For example in the drawing of the open U-beam (Figure 70) the sketched shear stress acts in the cross-section. However it induces a torsion moment around the centroid of the cross-sectional area and the longitudinal beam axis, which unfortunately results in a twisting of the cross-section and a lateral migration of the beam. This effect can be compensated by shifting the line of action of force F by d (see Fig. 70). We wish you a very pleasant day, and have a nice scientific trip!

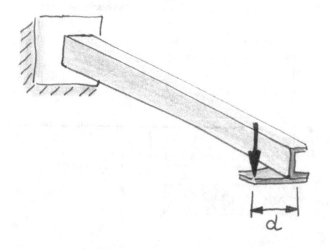

Figure 70: Skewed bending in non-skewed execution

2.6 The Frankfurter Formula

The moment has finally arrived when you can take the old frankfurter out of the fridge, where hopefully you have been storing it. We throw the pink thing into boiling water until it splits open. But look: The frankfurter has split lengthwise (in all other cases the sausage would probably be thicker than it is long – yuuuck. But even then, there is more tension in the direction of the width than axially.) All we need now is the right formula tool.

Take one frankfurter: The sausage skin splits with a hiss and shower of grease due to the expansion of its contents. In the field of boiler sciences this is referred to as a skin with interior pressure. Experience shows that the sausage splits lengthwise – if mommy/daddy have not taken the precaution of puncturing the rated splitting point with a fork.

Now to more technically sophisticated objects: Dr. Romberg's beer belly can be simplified as a boiler (wall thickness s, radius R, s<<R).[28] As a result of a beer pressure-refueling, his belly, or boiler, is pressurized with the internal pressure p_i. Something which is known to every backyard barbecue cook is referred to by mechanical engineers as follows:

The longitudinal stress, which by the way is

$$\sigma_z = \frac{R(p_i - p_a)}{2s} \quad,$$

is half as great as the circumferential stress: $\sigma_u = 2\,\sigma_z$

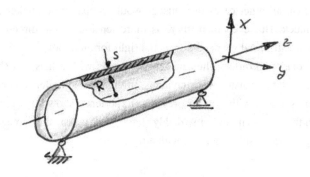

Figure 71: Model of the boiler

Dr. Romberg's demise due to continual pressure refuelings will be signaled either by liver failure or a longitudinal split of the abdominal wall. That's all, folks!

---

[28] However, the precondition of some sort of "longitudinal direction" similar to that of a boiler for Dr. Romberg's gut is a very bold assumption indeed! The following formulas are actually invalid for spherical bodies!

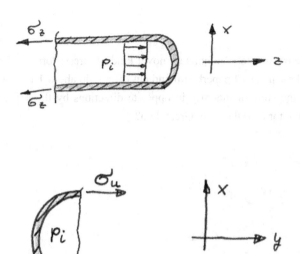

Figure 72: Stresses in Dr. Romberg's abdominal (boiler) wall

2.7 Torsion

Now we must turn to a completely novel kind of stress: torsion. To illustrate this, we take a piece of paperboard and cut off a strip about 1 inch wide. If we now twist the ends of the strip in opposite directions by the angle $\Delta\varphi$, well, then we have torqued the strip. Great, huh?

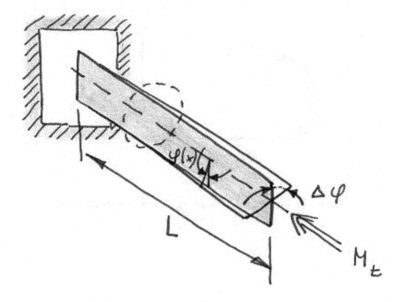

Figure 73: Torsion

As in the case of our bending beam, we can model a small element of the paperboard which we can use to study the process of torsion.

Even in this small sketch you can see that the springs in the cross-sections must be deflected laterally. The effect of torsion is thus a cross-sectional shear stress, which increases outwardly from the neutral fiber ($\tau=0$). The twist of the cross-sections caused by a torsional moment depends on the following variables:

- the stiffness of the model's springs, thus a variable similar to the modulus of elasticity E, but now for shearing action: shear modulus G,

- the distance of the springs, and thus of the sectional areas from the neutral fiber. Analogous to the model for the geometrical moment of inertia for bending, here we can also define a torsional geometrical moment of inertia I_t, which is also called "polar moment of inertia of an area",

- the length of the support.

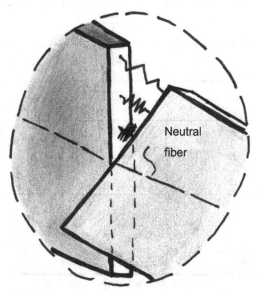

Figure 74: Elemental model for supports under a torsional load

As you probably know, you can solve any problem if you have the right formula and know the correct way to use it.

Here are some formulas for torsion:

129

	circular section (exact solution)	closed thin-walled section (approximate solution)	open thin-walled section (approximate solution)
example with associated stress distribution		 (n piecewise-constant wall thicknesses)	 (n piecewise-constant wall thicknesses)
maximum shear stress	$\tau_{max} = \dfrac{M_t}{W_t}$		
where	$r = R_a$	$b(s) \Longrightarrow$ min.	where $b(s) \Longrightarrow$ max.
twist ϑ	$\vartheta = \dfrac{d\varphi}{dx} = \dfrac{M_t}{GI_t}$		
torsion $\Delta\varphi$	$\Delta\varphi = \displaystyle\int_0^L \vartheta(x)\,dx$		
cross sections, M_t, G=const.	$\Delta\varphi = \dfrac{M_t L}{GI_t}$		
geometrical moment of inertia I_t	$I_t = \dfrac{\pi}{2}(R_a^4 - R_i^4)$	$I_t = \dfrac{4A_m^2}{\sum\limits_{i=1}^{n} \dfrac{s_i}{b_i}}$ (2. Bredt's formula)	$I_t = \dfrac{1}{3}\sum\limits_{i=1}^{n} b_i^3 s_i$
moment of resistance W_t	$W_t = \dfrac{I_t}{R_a} = \dfrac{\pi}{2R_a}(R_a^4 - R_i^4)$	$W_t = 2\,A_m\,b_{min}$ (1. Bredt's formula)	$W_t = I_t / b_{max}$ $= \dfrac{1}{3b_{max}}\sum\limits_{i=1}^{n} b_i^3 s_i$

(A_m: Area enclosed by the center line)

Geometrical moments of inertia and moments of resistance

We can find everything our heart desires in this table (as far as torsion is concerned). But since it is not entirely self-explanatory, here's an example for each case:

A built-in support with the circular cross-section (length L, shear modulus G, radius r) is loaded by torsional moment M_t in another direction than the plank. Determine the torsion $\Delta\varphi$ of the beam end.

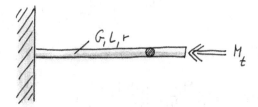

Figure 75: Torsion of a beam. Given: L, G, M_t, r.

First we have to determine the torsional geometrical moment of inertia I_t:

$$I_t = \pi\, r^4/2 \quad .$$

The torsion is then found according to

$$\Delta\varphi = \frac{M_t L}{G I_t} = \frac{2 M_t L}{G\pi r^4} \quad .$$

Now we do the same for the sketched thin-walled box girder with a rectangular cross section (width B, height H, wall thickness t_B, t_H):

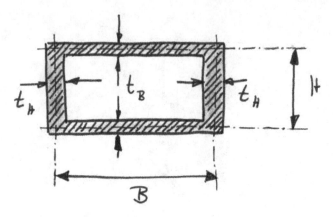

Figure 76: Hollow rectangular section

We'll use "the second board" here! For the torsional geometrical moment of inertia we obtain

$$I_t = 4A_m^2 / \sum_{i=1}^{4} \frac{s_i}{b_i} = \frac{4\,(BH)^2}{2\left(\dfrac{B}{t_B}\right) + 2\left(\dfrac{H}{t_H}\right)} \; .$$

Torsion is then

$$\Delta\varphi = \frac{M_t L}{G\,2B^2 H^2}\left[\left(\frac{B}{t_B}\right) + \left(\frac{H}{t_H}\right)\right] \; .$$

Now let's look at a third section: a thin-walled circular section, which is not slit in case a) and slit in case b). Now the prize question: Which section will be twisted more as a result of the torsional moment?

The practical-minded among you will probably guess that the slit section exhibits a "softer behavior". Well, let's see:

For the closed thin-walled pipe, the table gives us

$$I_{t,a} = \frac{4(\pi R_M^2)^2}{\dfrac{2\pi R_M}{t}} = 2\pi t R_M^3 \; ,$$

the slit pipe leads to

132

$$I_{t,b} = \frac{1}{3} t^3 2\pi R_M$$

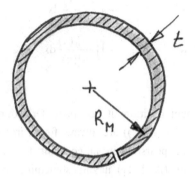

Figure 77: Slit circular section

The torsion thus results in

$$\Delta\varphi_a = \frac{M_t L}{2\pi G t R_M^3} \quad \text{resp.} \quad \Delta\varphi_b = \frac{3 M_t L}{2\pi G t^3 R_M}$$

Also interesting is the result for the ratio of displacements:

$$\frac{\Delta\varphi_a}{\Delta\varphi_b} = \frac{t^2}{3 R_M^2} \ll 1, \quad \text{since thin-walled section, } t \ll R_M'.$$

Our calculations thus confirm the practical-minded approach mentioned above. And that's just about it as far as torsion is concerned!

After this exhausting ascent across rugged terrain to the heights of the strength of materials, we would now like to take a well-deserved break – and with Dr. Hinrichs not wishing to let the time go by without being put to good use, he will cast a sweeping look at what we have reached so far. Here he is greeted by the following overwhelming panorama:

Elongation: $\quad \Delta L \quad = \quad \int \dfrac{N(x)}{EA(x)} dx$

133

$$\text{Torsion:} \quad \Delta\varphi \quad = \quad \int \frac{M_T(x)}{GI_T(x)} dx$$

$$\text{Bending:} \quad w' \quad = \quad -\int \frac{M_B(x)}{EI(x)} dx$$

$$w \quad = \quad -\iint \frac{M_B(x)}{EI(x)} dx$$

From such a paramount position, it occurs to Dr. Hinrichs that the stress variables ($N(x)$, $M_T(x)$ or $M_B(x)$) are always found in the numerator. In the denominator there first appears a material constant (E, G), closely followed by a characteristic quantity (A, I, I_T) having something to do with the cross-sectional geometry of the maltreated rope, rod or beam.

But for now we would to leave it at that and turn to a problem of stability.

2.8 Buckle Up

Let's take our elastic ruler, rest one end on the top of the table and load the other end by pressing it down with a finger.

If the ruler has been manufactured with 100% precision, the table is perfectly smooth and level, and we are able to exert 105% of the force precisely along the axis of the ruler... we could then keep pressing the ruler into the table without it having a sharpened point.

But general experience shows that the ruler will "bow" before that happens. If the ruler is still straight under a small load, we can still make it assume its bowed position by snapping a finger against the middle of the ruler without changing the exerted load. Ordinarily we cannot restore the bowed ruler to its original straight position by lightly snapping it. The seasoned mechanics veteran also refers to this as a stability problem. We have three possible positions of the loaded ruler:

- a straight ruler under a pressure load. This state is unstable, as we can readily determine by lightly snapping it with our finger,
- a ruler that is bowed to one side (either to the right or left).

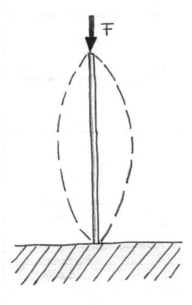

Figure 78: Buckling

These states are considered stable, since even after such a state is disrupted (with a light snap of the finger) it can be restored once more.

One usually tries to avoid this rather unsightly buckled state when dimensioning a support under this kind of load. Mechanics have been grappling with the calculation of this phenomenon for quite some time. The following table is the result of their sleepless nights and provides you with everything you need for buckling. However, the hard part in solving problems concerning buckling is to pick out the correct type of load. Just to test you: What type of load is involved in the case of the leg of the chair you are sitting on? You see, here we go again. Let's say we first decided on case 3. To simplify things, let's imagine a very soft chair leg made of rubber. The point of contact of the chair leg on the floor is the roller support as long as the chair leg remains in contact with its original position during buckling!

135

	Type of load	Buckling load F_{crit}
1		$\dfrac{EI\pi^2}{4L^2}$
2		$\dfrac{EI\pi^2}{L^2}$
3		$2.0457\dfrac{EI\pi^2}{L^2}$
4		$4\dfrac{EI\pi^2}{L^2}$

The ability of the roller support to slide suggests that during deformation the distance between the seat and the floor gets smaller. If the chair leg, due to insufficient friction with the freshly polished floor, slips to the side or if the seat is displaced horizontally while the chair leg maintains its point of contact with the floor, then we're dealing with case 1. The chair leg has a built-in support at the seat, i. e. even when the leg buckles it emerges perpendicularly from the underside of the seat. But we do hope that at the moment you are not buckling under and that the legs of your chair, couch or bed will hold out at least until you have read the last chapter before the problem exercises. The next chapter is going to get pretty dynamic! Things will start to move, rotate, slide and roll without skidding until they too get lost in boring equations – with the authors unable to do anything about it... or can they?... We'll see...

3. Everything in Motion: Kinematics and Kinetics

"The more you know, the more you know what you don't know!" [Loose translation of Confucius]

The by now fairly clever reader of chapters 1 and 2 can already do some pretty cool things like calculating the forces in rigid bodies that move with a constant velocity. As well as the strain that should occur at some point after the introduction of a load – only, of course, in the linear, elastic domain.

In order to inflate our gigantic understanding of the world even more we now want to turn our attention in the following sections to the processes of motion. So as not to exaggerate their general validity: We will observe only the uniform, accelerated or delayed motions of rigid bodies.

Let's take a rollercoaster ride as an example (Dr. Romberg knows exactly what he is talking about here, following the excessive enjoyment of beer again and again in his "own" room at night):

The breakneck ride begins (1) with the initial ascent, during which the coaster and its riders are raised to the maximum height. Then comes the most exciting part of the whole ride: The coaster moves rapidly downhill, the pulse races, Dr. Hinrichs starts to scream in front of us... almost a free fall (2)! Then it goes directly into the loop (3). Finally, we close our eyes and wait until the coaster slowly taxis to a standstill on the final even stretch (4). Now for the high point:

We're not gonna say: "Wheee..., great..., one more time, honey?" No, no, no – we'll spare ourselves such emotional outbreaks and try instead to evaluate the ride with our hip expressions. "Oh yeah! That's pretty cool! Heh heh. Heh heh heh." [Butthead]

3.1 Kinematics

(This comes from Kine: **K**nowing **I**s **N**ot **E**verything...)
First we have to determine the respective position on the rollercoaster. Various descriptions for this can be found in everyday speech:

"Yiiiiikkkessss, whoooaaahhhhhh!"

Here the scientist chooses the height H above the ticket stand as a coordinate. A further determination of position could sound like this:

"I could puke already, but we're only halfway through..."

Now we'll choose the path coordinate s (the already traveled part). Or does the sufferer mean one half of the ride time? Doesn't matter – we could also use this as a description. Let's call it t as in time.

In the curve we can choose – pretty complicated – the angle φ as the midpoint of the curve as a coordinate. The different types of descriptions and the used variables are dependent on each other: For example, the traveled arc s can be calculated with s = φ R. So if, let's say, a bicycle tire (radius R) has turned three times (3 x 360° ≅ 3 x 2 π), then we've covered the distance 3 x 2π x R.

To make things even more difficult, there is often some wild mixing going on between the different coordinates a, H, φ, s, t and the derivations of time, e.g. the velocity v = ṡ and the acceleration a = s̈. This obvious course of action – designating the position with the help of any given coordinate – is mysteriously referred to as kinematics. More official definitions sound like this:

Kinematics is the science of geometric and chronological motions

What's important here is that in all reflections on kinematics, no knowledge of forces is necessary. So now all statics and strength of materials drop-outs can attempt a grand comeback!!!!

Very smart mechanics switch over to polar- or cylinder coordinates in daring maneuvers – this is said to simplify the whole thing drastically.

We prefer polar co-ordinates!

But let's first of all determine the different descriptive quantities:

3.1.1. The "Who is Who" of Kinematics: Variables of Description

The height H doesn't present us with any problems – pretty much everyone knows how to deal with it (except for Dr. Romberg, who's not always at the height of information). The same goes for the distance x, which we'll use to describe the traveled distance of a rectilinear motion. "As the crow flies", so to speak. This type of motion is known as *translational*.

We still need the opposite of a translational motion: a rotation. We're now talking about a *rotary* motion, pretty logical, huh? We'll use the angle of rotation φ[30] to describe it.

To describe a general motion we need coordinates for the translation (x, y, H, s,...) as well as for the description of the rotation (point / central axis around which is rotated, angle of rotation φ, and radius R).

Is this enough info to clearly describe the motion?
Let us just for a moment engage in a conversation with two tanned buff guys at the bar:

"My car went 58.3 kilometers in 17.5 minutes!" "Then you're the one who was holding up traffic... I did 436,6 kilometers in 2 hours und 11 minutes." A man has to ask himself... who was faster? And yes, this conversation usually goes like this:

---

[30] Dr. Hinrichs insists on noting here that you can think of every rectilinear motion as a rotation around a point lying at an infinite distance in space – we give him our warm thanks for this important pointer!

"My car does 200 kilometers an hour." "Hey, mine too. So you bought the new *Audi* too?" (Dr. Hinrichs, this attracts the moolah... !)

So you take a reduced quantity – the traveled distance per time –, the velocity v, unit m/s (you have to pronounce the letter v for approximately 10 seconds, then you'll know why it's used for velocity!). ((This sounds to Dr. Romberg more like a desperate motorized attempt to change the velocity from v = 0 to v > 0: "vowvowvowvowvowvow... huh?")).

And then you can boast about how quickly you recently sped from the traffic light. "I did zero to eighty in five seconds," for example. But when you want to trump that, you unfortunately read in the automotive catalogues only about the acceleration time from 0 to 100 kilometers an hour (but what for, really – when the traffic light turns green? On the freeway after the traffic jam ends?). The scientific solution to this dilemma is supplied here by the acceleration a, which as the reduced quantity represents the change in velocity per time, unit $\Delta v/t = m/s^2$ (the letter ahh describes the admiring exclamations of the pedestrians observing one of the above-mentioned acceleratory events).

The same thing surely goes for the angle of rotation... Here you have analogously the angular velocity $\dot{\varphi} = \omega$ or Ω (both a Greek w, the latter character is also used for a tragic ending, which is visited upon many students of mechanics) and the angular acceleration $\ddot{\varphi}$.

3.1.2 Some Examples of Kinematics

We can practice making connections between the separate coordinates (and their derivatives) with the help of a few very intricate examples:

1) A cylinder rolling on a stationary base without slipping (chosen coordinates: motion x of the center of gravity and angle of rotation φ of the cylinder).

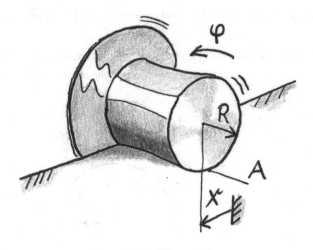

Figure 79: Rolling cylinder

There are some pretty sneaky considerations involved here. The cylinder touches the stationary base on line A. If the cylinder adheres to the base – i.e. rolls without slipping, then the cylinder must possess a velocity of zero at the point of contact. Tricky, huh?

If we look at a snapshot of the rolling, non-slipping cylinder at any given point in time, the cylinder moves around this point A. So the center of gravity moves at this particular point in time in circular path around point A! Whoever doesn't want to believe this should go back to 1) [see above] or just memorize the following sentence:

If a body rolls without slipping, then the point of contact of the body with the (stationary) base is at rest and the center of the body moves (for this given moment) in an orbit around the point of contact!

Of course, this applies *only to the exact moment* of our observation. However, in general it is the case that the roll path and thus the traveled distance of the cylinder is x = R φ. So x = R φ is also the arc through the body's center and the point of contact. The same goes for $\dot{x} = R\dot{\varphi}$ as well, naturally. For the ~~very stupid~~ novices: The point above a coordinate signifies the temporal derivative. So once more:

143

path of rotation (=x) = angle of rotation (=φ) × radius of rotation (=R).

2) The rolling cylinder on a belt (chosen coordinates: motion x of the center of gravity, angle of rotation φ of the cylinder and the motion y of the belt)

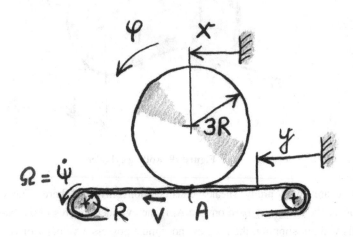

Figure 80: Rolling cylinder on a belt

For the driving wheels we have indicated the rotation with a ω for the angular velocity. But let's begin first with the angle of rotation ψ. If we turn the driving wheel by angle ψ, the belt will move by y = ψR. Then, if we establish the time derivative of this relationship, we get $\dot{y} = R\dot{\psi}$. But the derivative of the angle of rotation results in the angular velocity, $\dot{\psi} = \Omega$, so that you end up with $\dot{y} = R\Omega$. Now, in order to bring the rolling cylinder into play, it's best to use a trick: First, we'll look at two special cases where we'll separate the obviously different motions from one another:

    a) The cylinder moves with the belt and doesn't rotate ($\varphi=\dot{\varphi}=0$). In this special case it's immediately clear that $\dot{x} = \dot{y} = R\Omega$ has to apply to the velocities.

    b) The belt doesn't move, but the cylinder rolls ($\dot{y} = R\Omega = 0$, $\dot{\varphi} \neq 0$). This was the case in example 1). The result is $\dot{x} = 3R\dot{\varphi}$. So in both cases together we end up with...
    $\dot{x} = \dot{y} + 3R\dot{\varphi} = R(\Omega + 3\dot{\varphi})$.

3) The cable pull:

A block hangs on the pulley (Figure 81) with mass m. The ends of the cable are moved by x und y. Calculate φ and z.

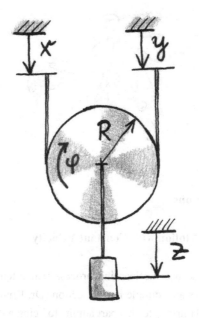

Figure 81: Cable pull

Perhaps you see the kinematical relationship $z = (x+y)/2$ right off the bat here. But let's take a more formal approach: We'll divide the motion once again into two special cases:

a) $x = 0$. What we've got here is the case of a rolling wheel again. The wheel rolls without slipping down thread x. In this case, $z = R\varphi$, $y = 2R\varphi$, so $z = y/2$.

b) $y = 0$. Now the whole thing in reverse. $z = -R\varphi$, $x = -2R\varphi$, so $z = x/2$.

The overlay of both types of motions results in $z = (x+y)/2$ and $\varphi = (y-x)/2R$. This is how sneaky kinematics can be!

145

One puke,
mass m

3.1.3 Special Motions

3.1.3.1. Circular Motion with Constant Velocity

We're not going to use the learning process that often causes people to run around in circles as an example in this section (Dr. Romberg knows about that well enough; Dr. Hinrichs for his part admits to being a disciple of translational learning).

Instead, we'll go back to our roller-coaster and observe the depicted loop. Let's do something completely different here and make an assumption that has nothing to do with reality. We'll assume that we move through the loop at a constant velocity v. In the acceleratory thrill of a luxury coaster with a couple of loops you will notice that given the same beginning velocity, it takes much longer to go through the bigger loop than through the smaller one. In both cases, an angle of 360° around the center of the loop must be traveled. So this must mean that the angular velocity ω for the bigger loop has to be smaller, right? We can sum up this correlation with the following formula:

$$v = \omega R$$

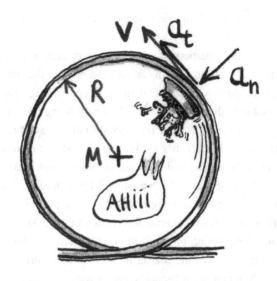

Figure 82: Acceleration during a circular ride

If we remove a part of the loop's track – without being inhabitants of the car ourselves – then the next set of riders will of course experience a slightly different ride route: Without that piece of track the car won't follow the loop, but will instead be tangentially shot out of the curve.

On the other hand, you could naturally deduce from this purely academic (!) consideration that with the presence of a track, the path of the car must be "bent" by the track in the direction of the circular path. In other words, the trajectory velocity of the car takes a different direction at every point of the loop. So the direction of the velocity vector, which is always tangentially aligned with the circular motion, is constantly changing. And this "bending" leads at every point in time to a *radial velocity change* a_n, also known as "centripetal acceleration". *At every point in time the radial velocity component of the car changes to the direction of the center of the circle* by the change in velocity (acceleration) a_n. This centripetal acceleration amounts to

$$a_n = \frac{v^2}{R} = R\omega^2,$$

and as such is quadratically dependent on the velocity at which the loop is traveled.

But now, the question of centrifugal acceleration or centrifugal force directly imposes itself on us. This is what the man on the street calls the force that presses us down onto the seat in the loop. Strictly speaking, we aren't being pressed onto the seat, but rather the seat is being taken under our "seats" in a different direction. But our mass would rather stay at rest or complete a uniform motion. *It is inert.* It's the turning around that requires the force that we then feel. If the coaster is thrown out of the loop and flies straight, then this force suddenly disappears. So the source of this force is the inertia of our carcass. Dr. Hinrichs likes to refer to this phenomenon as the county-fair-paradox. (For more on such inertial forces see chapter 3.3)

So that's everything on uniform circular motion..., but of course you'll complain right away that the coaster in the loop is not driven and, for reasons unknown up to this page, it becomes slower with increased height in the loop. Correct, correct! So now we come to general circular motion:

3.1.3.2 Circular Motion with Variable Velocity

In the case of circular motion with variable velocity the coaster still has to be accelerated in the direction of the center of the circle, naturally, so that it doesn't break out of the course. So the following still applies:

$$v = \omega\, R, \qquad a_n = \frac{v^2}{R} = R\omega^2.$$

But when the velocity of the coaster changes, it results in a change in the angular velocity ω as well. So, in addition, the coaster is accelerated or slowed tangentially to the orbit. The tangential acceleration a_t amounts to

$$a_t = \dot{v} = R\, \ddot{\varphi} = R\, \dot{\omega}.$$

Both of these acceleratory components can be added together vectorially and thus give us the resultant acceleration a:

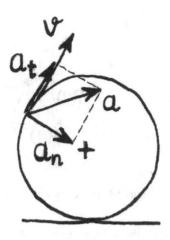

Figure 83: Acceleration during a circular run

3.1.4 The Velocity Pole

It really doesn't matter for our riding pleasure whether we move in the coaster through the loop or whether we weld the coaster to a Ferris wheel with the radius of the loop and propel the Ferris wheel with the calculated angular velocity ω. So you can regard the motion in the loop as a rotation around the axis of the Ferris wheel or the central point of the loop. This applies universally to any given motion:

Every motion can be understood as a rotation around a point.

Of course, the central point of the outlined general motion is not fixed like on the Ferris wheel, but rather moves along a curve. And the corresponding radius R(t) is also a function of time. As described earlier, the same motion is produced at the moment of observation when we weld the moving body onto the edge of a disk with radius R and central point M. But the velocity of a circular element decreases linearly in the direction of the center of the circle as outlined.

149

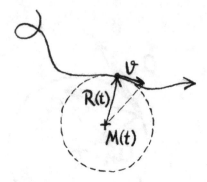

Figure 84: Trajectory with velocity pole

In the center of the disk the velocity is zero. We refer to this point in the following sections as the velocity pole of motion!

150

A velocity pole of a motion is the point in a body that is at rest for the moment of our observation and around which the body rotates at this particular moment.

As in our example with the rollercoaster, this velocity pole doesn't have to be a point within the moving body.
Let's look now at a random body that rotates around its velocity pole Q:

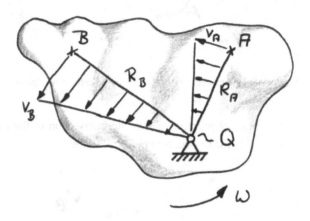

Figure 85: Velocity profile of a rotating rigid body

We know from the circular motion that the velocities of the body points A and B result in the following:

$$v_A = R_A\,\omega, \qquad\qquad v_B = R_B\,\omega \qquad .$$

Furthermore, what is striking here is that all velocities of the body stand vertically on the connecting lines AQ and BQ. This must be the case, since – if it weren't – the distance from A to Q would change. And this would mean that the body would be torn apart.
But the problem is often the reverse: For a predetermined course like that of our rollercoaster, for example, it is the velocity pole Q(t) that is sought.

Let's throw together some recipes from the previous deliberations:
Construction of the velocity pole

151

a) from two given body points A and B with the velocity directions:

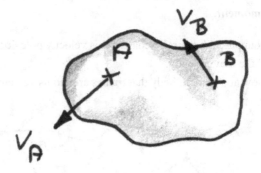

Figure 86: Determination of the velocity pole from two known velocities

- The velocity pole emerges as the intersection of the straight lines vertical to the velocities in points A and B:

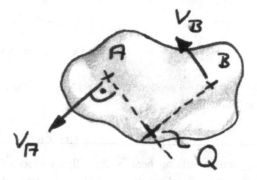

Figure 87: Determination of the velocity pole from two known velocities

Here we have a special case: the straight lines are parallel, i.e. they intersect in infinity. However, this means that in this borderline case you don't have a rotation, but rather a translation.

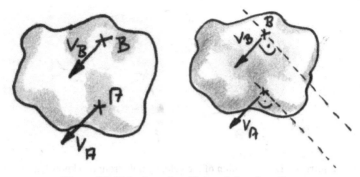

Figure 88: Velocity pole with translational motion

b) from the amounts of two velocities v_A and v_B, when both stand vertically on the connecting line of the corresponding points A and B.

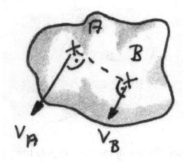

Figure 89: Determination of the velocity pole from two known absolute values of velocity

- The velocity pole emerges as the intersection of the connecting lines of the points A and B and of the apexes of the arrowheads of the vectors.

153

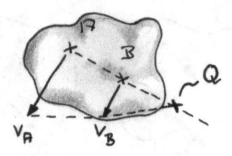

Figure 90: Determination of the velocity pole from two known absolute values of velocity

c)   from the given velocity v_A and the angular velocity ω in the body point A

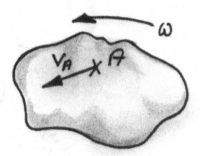

Figure 91: Determination of the velocity pole from a velocity and the angular velocity

- The velocity pole lies on the perpendicular to the velocity vector through point A; the distance R is determined by R=v/ω.

Note: The direction in which we have to move from A on the perpendicular clearly emerges from the given ω!

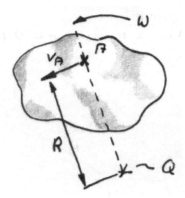

Figure 92: Determination of the velocity pole from a velocity and the angular velocity

Here's an example:

The sketched driving crank 1 turns in the position shown at the angular velocity Ω.

a) Determine the velocity poles of rods 1 through 3! (That's an order!)

b) Can you determine the angular velocities of rods 2 and 3 in the position shown?

(No! And thus I have answered the question *absolutely correctly*...)

c) For which position are the angular velocities of rods 2 und 3 the same?

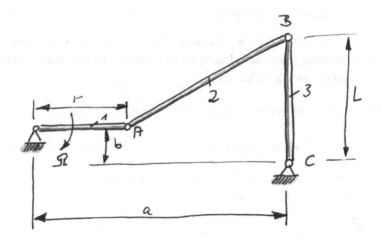

Given: Ω, c, L= 4c, a=10c, b=2c, r=3c

a) From rules a) – c) the velocity poles Q_1, Q_2 and Q_3 result:

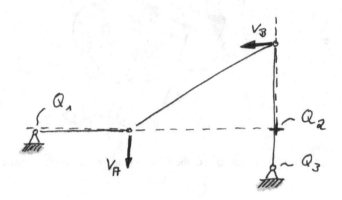

b) For velocity v_A and rod 1 the following applies: $v_A = \Omega\, r$. The same thing naturally goes for rod 2, so that the joint doesn't fly apart:

$$v_A = \omega_2\,(a{-}r)$$

$$\Rightarrow \quad \omega_2 = r\,\Omega/(a{-}r) = 3\Omega/7$$

You do the same for rod 3, and end up with:

$$v_B = L\,\omega_3 = (L{-}B)\,\omega_2, \qquad \Rightarrow \quad \omega_3 \quad = \quad 3\,\Omega/14$$

So the point here is that you often equate the velocities of two components on the connecting joint and calculate these velocities in two different ways by means of two angular velocities!

c) Yep, a little playing around... and voilà: There's the solution...

(Dr. Hinrichs maintains that it also has to work without the "playing around"...)

What we've completely neglected up to now is the dependence of the variables x, v, a, H, φ, $\dot{\varphi}$, $\ddot{\varphi}$... on time.

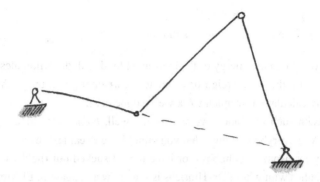

To put it more clearly: All of the deliberations in the thick pile of pages that you've worked through already relate to just one moment in time – a snapshot. But now for the fun stuff. We are going to turn our attention towards motion within a longer period of time. Our long-term goal will be to also consider the forces that lead to the motions, and influence them. And so we come to

K   i   n   e   t   i   c   s
!

3.2   Kinetics

3.2.1   The Energy Theorem

The energy theorem is really just common knowledge in scientific packaging: Everything that goes in must come out. Everyone knows about this from observing their wallets.

The same balance applies for our nutrition: Everything that we consume in the course of a week

a) is transformed into energy, i.e. physical activity or simply warmth,[31]

---

[31] Dr. Hinrichs interjects that it doesn't necessarily have to do with sports.

b) widens our waists,[32]

c) is excreted in fluid or solid form.[33]

So if we add up the energy of the consumed food with the Kilojoules like when you're on a diet, and weigh or estimate the amount given by c), then we can actually calculate how much work we've done in a)!

With such a balance, we can then... well, *balance* for longer periods of time! And everybody knows that you either have to eat less, exercise more (not exactly Dr. Rombergs hobby), or have the fat sucked out (hobby of the face-lifted ladies with whom Dr. Hinrichs is seen now and again in El Arenal).

Now, on to the processes of motion: Everything that goes into our animated bodies has to stay somewhere or come out again. An example: If we throw a ball into a sandbag, then the energy of the throw is completely transformed into the deformation of the sandbag. But if the ball lands on the lawn at a sporting event after flying for 20 meters, then the energy of the throw will cause a small crater at first. The residual energy is then used up by friction and the deformation of blades of grass while the ball is rolling over the lawn. The sum of the deformation energy (crater, blades of grass) and the energy consumed by the frictional processes during the roll results in the original energy of the throw (Dr. Hinrichs also makes reference to the air resistance).

So, the total energy remains constant during the entire process of motion. But it doesn't matter to the crater at all whether it is created by the ball's fall from a tower with height H (meaning the result of so-called potential energy in the earth's gravitational field), by the launching of the ball with a slingshot (the potential energy of a spring) or by the initial speed of the throw (kinetic energy).

In order to be able to make calculations at all with such a balance, you first have to set up a table with the energies in the most varied forms (kind of like the dietary table where for each type of food the number of calories is indicated, that those meandering through a diet excrete). Voilà, here it is:

---

[32] Dr. Hinrichs interjects that it's not only the waist that is widened.

[33] Dr. Romberg interjects that some excretions can also be in gas form.

Type	Cause		Symbol	Calculation
	general			$E = W = \int F dx$
potential energy: $E_{pot} = U$	special cases	gravitational field here: Choose the proper reference level $(E_{pot} = 0)$!		$E_{pot} = U = mgH$
		spring, elastic deformations		$E_{pot} = U = \dfrac{1}{2} c\, x^2$, torsion spring: $E_{pot} = U = \dfrac{1}{2} c_\varphi\, \varphi^2$
		gravitation		$E_{pot} = U = \Gamma \dfrac{Mm}{r}$ mit $\Gamma = 6{,}672 \cdot 10^{-11}$ $[m^3/(kg\, s^2)]$
	pure translation (linear motion)			$E_{kin} = T = \dfrac{1}{2} m\, \dot{x}^2$

kinetic energy: $E_{kin} = T$	pure rotation			$E_{kin} = T$ $= \frac{1}{2} J^Q \dot{\varphi}^2$
	translation and rotation			$E_{kin} = T =$ $\frac{1}{2} J^Q \dot{\varphi}^2$ **or** $E_{kin} =$ $\frac{1}{2} J^C \dot{\varphi}^2 + \frac{1}{2} m \dot{x}^2$
consumed (dissipated) energy: $E_{diss} < 0$	forces against the direction of motion	F=const., z. B. Coulombian friction with $F = \mu F_N$		$E_{diss} = F\, s$
		plastic deformations, $F = f(x)$		$E_{diss} = \int F dx$
input: $E_{zu} > 0$	Forces in the direction of motion			$E_{zu} = \int F dx$

The energy theorem for condition 1 and a later condition 2 would then be as follows:

$$E_{pot,1} + E_{kin,1} + [\, E_{diss} + E_{zu}\,] = E_{pot,2} + E_{kin,2} \,.$$

The energy changes given in brackets are the result of active forces during the motion from 1 to 2. The energies at moments 1 und 2 generally consist of potential and kinetic energy.

The infernal formula reads like this: The energy at moment 2 (index 2) corresponds to the energy at moment 1, increased by the input during the motion from 1 to 2. One should notice that energy is always extracted by the friction, i.e. $E_{diss} \leq 0$. Many problem formulations assume that during a motion no energy is consumed and added – an assumption that would support the notion that there is a perpetual motion machine!!!

Whoever thinks about this knows that there isn't such a thing. Such systems are referred to as „conservative" (lat.: "to not think"). In such cases, you can use the following simplification of the energy theorem:

$$E_{pot,1} + E_{kin,1} = E_{pot,2} + E_{kin,2} = E_{ges} = const .$$

It's often simply written like this:

$$T_1 + U_1 = T_2 + U_2 \quad ,$$

where the T's represent the kinetic and the U's the potential energy. The subscripts also refer here to before (1) and after (2*). With potential energy it is very important to choose an appropriate zero level.* It usually makes sense here to „set to zero" one of the two expressions for potential energy (U).

Even though almost all of us know (except for Dr. Romberg, who hasn't yet completed research on the subject) that there is no perpetual motion machine, mechanics experts maintain that one type exists, since the small addition can be found „frictionless" or undamped in the problem formulation. For this special case, the total energy of the system E_{ges} remains constant. We will now turn our attention towards a typical application:

3.2.1.1 The Matter of the Free Fall

Let's imagine the Starship Enterprise as it flies at warp 23 through the galaxy. As a result of zero gravity and the vacuum no propulsion is needed once „cruising speed" has been reached and Captain James T. Kirk doesn't have to register any slowing meteorite showers (Dr. Romberg points out that the danger of a wormhole does exist here) into the logbook. The trip would lead more or less straight into Nirvana until the end of time (a further remark by Dr. Romberg: In his opinion – from which Dr. Hinrichs would like to emphatically distance himself at this point until proof of the opposite has been established – space is temporally warped).[34]

It looks different if, in terrestrial regions, one chucks a body with an initial speed of v_1 from the height $y = y_1$ in the direction of the ground or sky (condition 1). As is generally known, these flying objects usually detonate after finite time with a velocity of v_2 on the ground again (height $y_2 = 0$, condition 2). Let's now examine this kind of flying motion more closely: We can very quickly derive the energies that come into play for this motion from our energy table: At the point of the throw (condition 1) kinetic energy results from the initial speed, and potential energy from the earth's gravitational field. From the energy theorem follows:

$$E_{ges} = E_{pot,1} + E_{kin,1} = m\,g\,y_1 + \frac{1}{2}\,m\,v_1{}^2 = E_{kin,2}$$

$$= m\,g\,y_2 + \frac{1}{2}\,m\,v_2^2 = \frac{1}{2}\,m\,v_2^2.$$

This formula applies whether or not the flying object is thrown up or down! But why? Because – when it's thrown up – the flying body also has to come down again at some point... and because of the conservation of energy, it goes past the place of the throw at the same velocity with which it was thrown.

So now we can calculate for every throw height y_1 above the ground the corresponding velocity or the maximum altitude: At the highest point of the flight path the velocity is zero – otherwise the object would keep on flying. We'll simply relocate condition 2 to the highest point of the flight path, so that $\dot{y}_2 = 0$ applies. Thus, according to the formula above, the following applies to the special case $y_1 = 0$ (i.e. we'll let the y-coordinate start counting at the place of the launch):

$$y_2 = y_{max} = H = \frac{v_1^2}{2g}\ .$$

---

[34] It would seem that the authors have touched a sore spot in the proof reader's past: Original sound byte: "The proof is being used today by anyone who has ever used a GPS-watch. With the satellites, the time dilatation has to be taken into consideration on account of the earth's mass(!), otherwise the location is incorrectly calculated." Dr. Romberg's answer: "You see, Dr. Hinrichs! ...But what does the faculty have to do with the earth's mass?"

And in the case of the bottle rocket every little boy will annoy the clever mommy or the clever daddy with the question of how long it takes for the rocket to arrive at the highest point. In order to answer that, we have to retreat (feign a cough attack or the need to relieve oneself) and make a few little calculations (here a little warning: This is going to be difficult, so leave the little boy outside!)

> *A little girl asks her dad, who has a doctorate in (mechanical) engineering:*
> *"Hey Dad, why didn't Beethoven finish his last symphony?"*
> *"Um.... not sure..."*
> *"Hey Dad, what do sociologists do?"*
> *"Um... let's see... you know, I don't know that either, I don't know any..."*
> *"Hey Dad, why are there sometimes so many locusts in Africa and sometimes not?"*
> *"Well... uh... I've heard of that ... dunno... but I must say: You ask really good questions. Keep on asking, because questions help you learn!"*

So we know that the total energy during the flight remains constant:

$$E_{ges} = m\,g\,y + \frac{1}{2}\,m\,\dot{y}^2 \qquad .$$

But if we derive this equation according to time (hey!), then follows

$$\frac{d}{dt}\,E_{ges} = 0 = m\,g\,\dot{y} + \frac{1}{2}\,2\,m\,\dot{y}\,\ddot{y} \text{ (chain rule differentiation).}$$

We divide this by $\dot{y}$ und convert:

$$\ddot{y} = -\,g \qquad .$$

It's crazy! We've found out that only the acceleration due to gravity g (usually g = 10 m/s² is enough, for Dr. Hinrichs: g = 9.81 m/s² (dependent on the location on earth and on the distance to the center of the earth)) affects the body in free fall (better: in „free ascent"). Great. But before we can let the screaming and door-scratching youngster back into the room again, we still have to figure out from the acceleration $\ddot{y}$ velocity $\dot{y}$ and the height coordinate y with dependence on time:

$$\dot{y}(t) = \int_0^t \ddot{y}(t)\, dt = -g\, t + v_0$$

with $v_0 :=$ initial speed for t=0

and $y(t) = \int_0^t \dot{y}(t)\, dt = -\frac{1}{2}\, g\, t^2 + v_0\, t + y_0$

with $y_0 :=$ launch height for t=0 .

Or, how aptly the ancient heroes of mechanics formulated it [8]:

> If a body is cast upwards, then the uniform gravity pours forces into it, and extracts from it velocities proportional to time. The time of ascent to the greatest height acts proportionally to the velocities to be extracted and those altitudes, like the velocities and time together, or they stand in double ratio to the velocities. The motion of a body, thrown along a straight line that must emanate from the launch, is added to the motion that arises from gravity.

Got it?

For the calculation of the flight time t_{max} up to height y_{max} you get from $\dot{y}(t_{max}) = 0$

$$t_{max} = \frac{v_0}{g}$$

Alright, let the little guy in... (Of course, if he asks about the initial speed of the bottle rocket, our condolences. And anyway, the rocket becomes faster and faster at the beginning, so an acceleratory force is involved, bad example!) At this point Dr. Romberg likes to show off his scars and tells a story from his childhood, when he developed a multi-stage rocket from two screwed-together bottle rockets connected to one another by a fuse – unfortunately, the second stage ignited only *after* the upper reversal point...

Now an example of the energy theorem:

In a shaft of height H a stone falls vertically downwards. After time T you hear a short suffocated scream at the shaft opening (velocity of sound c).

a) Determine height H!

b) How big is the relative error $\Delta H/H$, if you interpret time T as purely fall time?

Given: T = 15 s, g = 9.81 m/s², c = 330 m/s.

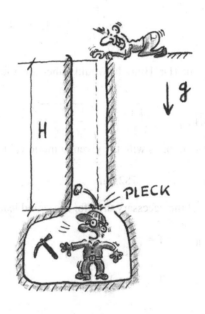

a)   The formula for the launch motion results in

$$x(t_{fall}) = H = \frac{1}{2} g\, t^2 + v_0\, t + x_0$$

with $v_0 = 0$ and $x_0 = 0$

$$\Longrightarrow \qquad t_{Fall} = \sqrt{\frac{2H}{g}}.$$

165

The sound spreads out by $x(t) = -c\,t$ from the impact surface, so

$$t_{sound} = \frac{H}{c} \ .$$

The total time T results from the sum of both times:

$$T = t_{fall} + t_{sound} \quad .$$

$$\Longrightarrow \qquad 15 = \sqrt{\frac{2H}{g}} + \frac{H}{c} \quad ,$$

$$\Longrightarrow \qquad H + c\sqrt{\frac{2}{g}}\,\sqrt{H} - 15\,c = 0 \quad .$$

With the p-q-formula (Dr. Hinrichs would rather use Vieta's theorem here) you get

$$\sqrt{H}_{1,2} = -c\sqrt{\frac{1}{2g}} \pm \sqrt{\frac{1}{2g}c^2 + 15\,c} \ ,$$

the only sensible solution is with the given numeric values:

$$H \quad = \quad 782{,}3 \text{ m} \quad .$$

b)   Disregard of the necessary time for the sound leads to

$$t_{fall} \quad = T = \sqrt{\frac{2H}{g}} \quad ,$$

so $H^* \quad = T^2\,g\,/\,2 = 1103{,}6$ m

$$\Delta H = H^* - H \ = 321{,}3 \text{ m}$$

$\Longrightarrow$ relative error: $\Delta H\,/\,H = 321{,}3\,/\,782{,}3 = 0.41$.

(Anyone who calculated a different relative error has to look for an absolute error in his/her solution.)

Debugging

We have foisted on you, as the basic assumption for what we've said here, that the throw or rather the free fall of our flying object always ensues nicely in the direction of acceleration due to gravity. But naturally, this restriction is just as unacceptable for all possible throw motions as if one were to open a zoo just for anteaters. We thus proceed now to the more general case, the so-called skewed throw.

3.2.1.2 Skewed Throw

Some famous heads have already wrestled with the problem of another special form of throw motion, a horizontal toss [8]:

> If an object 𝔄, by virtue of the throw motion alone, could at one given time follow the straight line 𝔄𝔅, and by virtue of the falling motion alone at the same time travel height 𝔄ℭ, then at the end of that time, if one completes the parallelogram 𝔄𝔅𝔇ℭ with combined motion, it will be at point 𝔇. Curve 𝔄ℭ𝔇, which it follows, is a parabola.

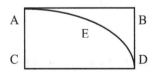

167

Although the facts have been made clear by this, here are some additional explanations:

With a horizontal throw from point A the thrown object will hit the ground just as fast at point D as when it's dropped from point A and lands at point C. So the general parabola of the throw is an overlap of the motions:

- horizontal motion with constant velocity v_x:

$$a_x = 0, \ v_x = \text{const}, \qquad x = v_x \, t$$

- vertical motion with acceleration as a result of the acceleration due to gravity:

$$a_y = -g, \qquad v_y = -g \, t, \qquad y = -\tfrac{1}{2} g \, t^2 .$$

Easy, huh? So the horizontal throw is described by the same equations in the same manner as the vertical throw – just through the addition of motion in a horizontal direction. You can imagine the whole thing as somewhat akin to the „Tom and Jerry" cartoons, where Jerry runs straight off of a cliff with a raging Tom breathing down his neck, stands still in the air, dares to look down and then – with horror in his eyes, falls vertically downwards: Reality is indeed different, but the result's the same (Dr. Romberg would have found the pair Wile E. Coyote and the Roadrunner (beep! beep!) more appropriate here).

Skewed litter

In the following sections, we'll occupy ourselves with a favorite free time occupation of mechanics: cherry pit tennis. Assuming you wanted to spit a cherry pit from height H to a goal at distance L at height h: how to do it, how much speed and which spitting slope α?

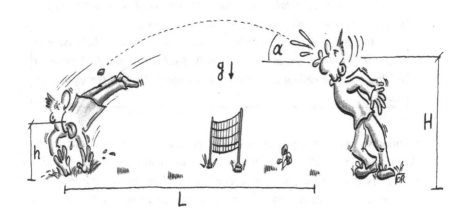

Here, too, we can first take apart the two motions:

The initial speed amounts to:

$$v_x = v_0 \cos\alpha \quad,$$

$$v_y = v_0 \sin\alpha \quad.$$

However, with these velocity components we can continue to separately calculate for the x- and y-directions in a familiar manner:

$$x(t) = v_x\, t = v_0 \cos\alpha\, t \quad,$$

$$y(t) = v_y\, t - 0.5\, g\, t^2 + H = v_0 \sin\alpha\, t - 0.5\, g\, t^2 + H\,.$$

At point in time t* the cherry pit should arrive at the goal. The following applies:

$$x(t^*) = L = v_0 \cos\alpha\, t^*,$$

$$y(t^*) = h = v_0 \sin\alpha\, t^* - 0.5\, g\, t^{*2} + H\,.$$

A manual laborer is locked in a cave for a week with a can of fish – after that week, the whole wall has been demolished by the throws of the workman – can open, manual laborer lives!

Then an engineer undergoes the same procedure – after the week's up, the whole wall is full of equations (throw parabola!), a part of the wall is damaged, can opened at the calculated spot, engineer lives.

And then the mathematician: After that week the whole cave is covered in writing, the mathematician is dead with a satisfied smile on his lips – on the ceiling is written: „Assumption: The can is supposed to be open ... ".

So now we have two equations with the unknowns α and v_0 at our disposal. We'll leave the solving of this rather mathematical problem up to the volunteers among you, who want to emulate Dr. Hinrichs. And for the first hot shots who send us correct answers, we're giving away an asbestos suit to their relatives. For all the others: The important thing here is the principle of overlap – and if you've gotten that much, the rest is a cinch!

And now another important trick: The choice of the coordinates' zero position is up to you. This applies for example to the zero position of x, y and t. With the right choice, you can save yourself a lot of work in the form of pages of algebraic reformulations. It doesn't make sense to begin the time axis with Dr. Romberg's loss of virginity; who knows when that was (or with that of Dr. Hinrichs; who knows when it will be). Similarly, it doesn't please the person asking if we begin at the North Pole when describing the way to the bakery – even though by so doing, we would present all the necessary information. ~~And in the same manner, certain textbooks seem incapable of making mechanics accessible. Sure, you can theoretically achieve this goal with them, but...~~

So pay attention:
Always at the starting - or target state
set the zero position of the coordinates!

For the last example this means:
Time begins counting at t = 0, when the cherry pit leaves the mouth. The coordinates $x(t=0) = 0$ and $y(t=0) = H - h$ describe the spitting position. This choice of the y-coordinate's zero position has the advantage of producing $y(t^*) = 0$ at the goal !!!

By the way – no mass turns up in these equations. So the calculated result is independent of what material the spat out object is made out of. Thus, if you don't have a cherry pit handy...

3.2.1.3 The Energy Theorem during Rotation

Here, a few more words on rotation: Until now, we have only calculated the kinetic energy for translational (a reminder: free of spin) motion with $E_{kin} = 0.5\ m\dot{x}^2$ (For those who don't know where the 0.5 comes from: It's another way of writing $\frac{1}{2}$ [or one-half]).

Unfortunately, it sometimes happens that a body rolls. In this case, the body doesn't just rotate – then it would stay in the same place. The rolling of a cylinder, for example, means rather a translational motion of the center of gravity and additional rotation. Let's begin first with pure rotation.

Following example:

Dr. Romberg (mass m_R: 72,5 kilograms + 2,5 kilograms emergency-liquor) sits at distance r from the rotational axis on a merry-go-round. After some hard propulsion work by Dr. Hinrichs (Dr. Romberg is athletic only in rare cases) he brings the merry-go-round to an angular velocity $\omega_{R,1}$.

Dr. Romberg leans back with obvious enjoyment in order to be alone with a bottle of beer at distance r (radius of merry-go-round) from the center of rotation. His kinetic energy (index R) is:

$$E_R = \frac{1}{2}\, m_R\, v_R^2 = \frac{1}{2}\, m_R\, \omega_{R,1}^2\, r^2\,,$$

Dr. Hinrichs just can't stop messing around with this equation and comes up with:

$$E_R = \frac{1}{2} m_R \, \omega_{R,1}^2 \, r^2 = \frac{1}{2} m_R \, r^2 \, \omega_{R,1}^2 = \frac{1}{2} J_R \, \omega_{R,1}^2 \quad ,$$

By doing so, he stumbles upon a new quantity, J_R, also called moment of inertia. You can't deny a certain analogy with the mass during translation. In the same way that mass represents a kind of resistance to motion, the moment of inertia represents a sort of resistance to rotation. The result of Dr. Hinrichs's messing around, namely

$$J_R = m_R \, r^2$$

shows in all its beauty what the moment of inertia is about, and that's first of all mass itself (otherwise there wouldn't be any inertia) and secondly on the distance of the mass to the rotational axis. But the latter also reveals something else: The moment of inertia is clearly dependent on the point of reference!
So now you calculate the energy E_R according to

$$E_R = \frac{1}{2} J_R \, \omega_{R,1}^2 \, ,$$

if you disregard the mass of the merry-go-round!!! This formula is set up pretty similarly to the one for translational motion.

1)   So the J, which is also referred to as moment of inertia, corresponds to mass m for the translational motion during rotation. Dr. Hinrichs will now say that in Dr. Romberg's example the inertia will be infinitely large[35].

It is essential for the observation of the energy, where the connoisseur has taken a seat: If he sits on the rotational axis, then he's very easily set into rotation (=>(moment of) inertia J small, energy small). But if he's hanging in a seat at distance $r = 1$ km from the rotational axis, then the merry-go-round will only be accelerated by a large expenditure of energy (=> moment of inertia J large, energy large). The moment of inertia then also depends, of course, on the

---

[35]Dr. Hinrichs also maintains, on the other hand, that Dr. Romberg is the only object in the entire universe that possesses a bit of mass, but no energy.

location of the center of rotation. *So the same mass can cause different moments of inertia.* The moment of inertia J_R depends – corresponding to the geometrical moment of inertia – quadratically on radius r:

$$J_R = (m_R + m_{liquor})\, r^2 .$$

2)   The angular velocity $\dot{\phi} = \omega$ for the rotation corresponds to the velocity for translational motion. This goes – corresponding to velocity with translational energy – quadratically into the calculation of energy. With that equation, Dr. Hinrichs can additionally calculate the energy that he's provided to Dr. Romberg.

Following this calculation time of about fifteen seconds, Dr. Hinrichs (mass M = 97,5 kilograms + 2,5 kilograms scientific literature) also gets onto the merry-go-round from a state of rest at distance r, because he – for scientific reasons, naturally – wants to experience the feeling of a centripetally accelerated body. And what happens?

Of course – the merry-go-round slows down. The reason for this is that Dr. Romberg's energy is now being used by both people (You could also say: Dr. Hinrichs is sponging on others). So energy E_1 divides itself up into E_R (Romberg) und E_H (Hinrichs):

$$E_1 \quad = E_2 \quad = E_R + E_H$$

$$= \frac{1}{2}\, J_R\, (\omega_{R,\,2})^2 + \frac{1}{2}\, J_H\, (\omega_{H,\,2})^2 .$$

As a compulsory condition for the fact that both are sitting on the same merry-go-round, the following still applies:

$$\omega_{R,\,2} = \omega_{H,\,2} = \omega_2 .$$

While Dr. Romberg occupies himself even more intensely with his next bottle of beer, Dr. Hinrichs calculates to what extent the velocity of the merry-go-round has decreased.

And now for a little alternative: Of course, as a *theoretician* you can assume that both of the academic bodies fuse together into one – naturally just theoretically. So you're really just acting as if the bodies become one with each other; the reality is obviously different. In this case we can formulate energy E_2 more simply:

$$E_2 = \frac{1}{2} J_G \omega_2^2 .$$

For the total moment of inertia of the fused bodies the following applies:

$$J_G = (m_R + m_H + m_S + m_L) \, r^2$$

One more important little note: It doesn't matter at all in looking at the energy whether:

a) The erudite scholars Romberg[36] and Hinrichs are sitting on the same seat on the merry-go-round or on opposite seats,

b) Dr. Hinrichs is sitting straight up at distance r from the rotational axis or is lying in an arc with radius r at a distance from the rotational axis.

In the following, a personal tragedy comes to pass for Dr. Romberg: As a result of his increasingly erratic grip, he loses a full bottle of beer in a radial direction. While Dr. Romberg is close to tears and comforts himself with the last remaining beer bottle, Dr. Hinrichs investigates the „Mechanics of Tragedy". He's especially interested in what happens to the angular velocity of the merry-go-round *shortly* after his "lapse" with the bottle. Does the angular velocity of the merry-go-round diminish, since energy is being taken away by the bottle flying away?

We already know from circular motion that the circumferential velocity is $v = \omega R$. At this velocity v the body is tangentially shot out of the circular path. Now the equations for the balance of energy pass before Dr. Hinrichs's mind's eye:

---

[36] Those "in the know" suspect that Dr. Romberg "acquired" his title through scurrilous channels.

$$E_2 \quad = \quad E_3$$

$$\tfrac{1}{2}\, J_G\, \omega_2{}^2 \quad = \quad \tfrac{1}{2}(J_G - m_{\text{beer bottle}}\, R^2)\, \omega_3{}^2 + \tfrac{1}{2}\, m_{\text{beer bottle}}\, v^2.$$

And when Dr. Hinrichs then plows through the equations, it turns out, strangely enough, that $\omega_2 = \omega_3$. But that's obvious:

For the state of motion in question, it doesn't matter whether the beer bottle in state 2, meaning before the tragedy, is connected to the mouth of Dr. Romberg – or, with no connection to the merry-go-round, floats on an invisible rod (at radius R with angular velocity ω_2) around the rotational axis in front of Dr. Romberg's red nose. But then the motion of the merry-go-round doesn't have to change, if at some point the bottle comes loose from the rod and shatters on the ground.

Or the other way around: Dr. Hinrichs got on the merry-go-round *from a state of rest*, and thus had to be supplied with energy by the merry-go-round, while the beer bottle does not change the state of motion or the energy.

So this is another example demonstrating the differing choices of frame of reference: We first viewed Dr. Romberg and Dr. Hinrichs as separate systems, then the fusing of both and finally the fusing without the beer bottle as one system and the beer bottle as a second system.

But now back to moment of inertia again. The distance of the mass from the rotational axis must play a decisive roll in the rotational energy. This dependence (and the chosen letter J) strongly allude to the geometrical moment of inertia in chapter 2.

In the following sections we will denote the moment of inertia of a body with J^Q relative to a (stationary) center of rotation Q. The center of rotation can always be found in the following by the J in the upper right corner, and the signs at bottom right denote the rotational axis. The unit for moment of inertia is [kg m^2].

For some standard objects the moment of inertia can also be represented in table form (see the table below) – but you'll search for the merry-go-round with the beer bottle in vain.

The most important moments of inertia:

cylinder & disk: (short solid cylinder, h=0, r=0) annulus: (short hollow cylinder, h=0, r=R)	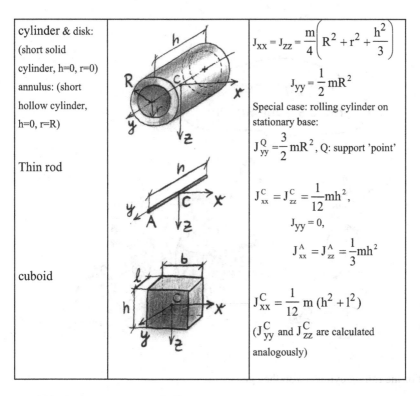	$J_{xx} = J_{zz} = \dfrac{m}{4}\left(R^2 + r^2 + \dfrac{h^2}{3}\right)$ $J_{yy} = \dfrac{1}{2} mR^2$ Special case: rolling cylinder on stationary base: $J_{yy}^{Q} = \dfrac{3}{2} mR^2$, Q: support 'point'
Thin rod		$J_{xx}^{C} = J_{zz}^{C} = \dfrac{1}{12} mh^2$, $J_{yy} = 0$, $J_{xx}^{A} = J_{zz}^{A} = \dfrac{1}{3} mh^2$
cuboid		$J_{xx}^{C} = \dfrac{1}{12}\, m\,(h^2 + l^2)$ (J_{yy}^{C} and J_{zz}^{C} are calculated analogously)

And if a body moves around all weird, e.g. rolls, then we can always ascertain the velocity pole of the body and then determine the kinetic energy for *pure rotation* around the velocity pole:

$$E_{kin} = \frac{1}{2} J^Q \omega^2 \quad .$$

But oh no! How do we calculate for the given moments of inertia the moments of inertia around a point displaced at distance d? (For example, the moment of inertia of a rolling cylinder relative to its point of support, i.e. to its velocity pole?)

As with the geometrical moment of inertia, Steiner's theorem aids us here!

$$J_{xx}^Q = J_{xx}^C + m\,d^2 \qquad .$$

Hey, Steiner! ... here's your share!

For the rolling cylinder you then get

$$J_{yy}^Q = J_{yy}^C + m\,R^2 = \frac{3}{2}\,m\,R^2.$$

(For practice, one might verify the result in the table for the rolling cylinder and the rod tilting around A and C!)

For practice, *you shall also verify* the result in the table for the rolling cylinder and the rod tilting around A and C!!!

Whew, that was hard stuff! For pedagogical and didactic reasons, Dr. Hinrichs has deemed it very worthwhile at this point to come up immediately with a nice example. Original sound byte: "That rounds off the matter amazingly!"

The problem stems from Dr. Romberg's past (don't worry, we're not going to get into the darkest chapters here). The object of examination will be

178

the college interfraternity games, at which Dr. Romberg was able to garner more success than was the case with his attempts at achieving academic success. Subject: "Extreme Tea Bag Pitching", which is a sport in East-Frisia, a widely unexplored coast area in Germany, where Dr. Hinrichs where found many years ago by a brave missionary.

Figure 93: Extreme Tea Bag Pitching

What is sought is the toss length x_{max} at given R = 1 m, ω = 1 /s, toss height H = 1.5 R, toss angle α = 45°, g = 10 [m/s²].

First, we set up the energy theorem for the moments directly before and after the toss of the tea bag, in order to figure out the toss velocity:

$$E_{sought} = \tfrac{1}{2} J \dot{\phi}^2 + mgH = \tfrac{1}{2} mR^2 \omega^2 + mgH \text{ (before)} = \text{(after)} \tfrac{1}{2} mv^2 + mgH \quad .$$

After extensive conversions we get v = Rω = 1 m/s. Great! We already knew that from circular motion. So, much ado about nothing.

Now, quickly, the toss equations (zero-point of the y-coordinate: goal ground, zero-point x-coordinate: place of toss, zero point time axis: toss, arrival on ground t*):

$$x(t^*) = v_x \, t^* = v \cos\alpha \, t^* = x_{max} \quad ,$$

$$y(t^*) = v_y \, t^* - 0.5 \, g \, t^{*2} + H = v \sin\alpha \, t^* - 0.5 \, g \, t^{*2} + H = 0 \quad .$$

179

Resolving results in ... $t^* = 0.623$ s and $x_{max} = 0.4405$ m.[37]

The next thing we want to do is to focus our attention a bit more on forces. And so we come to the principles of the illustrious Sir Isaac Newton:

3.3 Laws of Motion

In statics, we assumed that a body is „static", that is to say that it's at rest or moves at a constant velocity when the sum of the forces acting upon it is zero. It's also clear that – in the case of a force imbalance or a „force surplus"– there occurs a change in motion. Old Newton discovered the way in which these changes in motion can be quantified:

> **Change in motion is proportional to the impact of the moving force and occurs in the direction of the straight line upon which that force acts.**

Please memorize and never forget:

$$\Sigma F = m \, a = m \, \ddot{x} \text{ or } m \, \ddot{y} \qquad . \qquad \textbf{(Newtonian axiom)}^{38}$$

And here's another important tip from the old hand:

Be careful with the signs! Always pin down the coordinates x, y with directions, and then collect and sum up the forces moving in this direction with a positive sign and those going against these directions with a negative sign. If you end up with a positive sum, there will follow an acceleration in positive coordinate direction und vice versa!

---

[37] Dr. Romberg's commentary: "I threw further back then at my interfraternity games. My tea bag probably had less mass."
Oh geeeeeezzzz!!!!!!!! A request to the reader not to store this unqualified remark: Mass has nothing to do with it, Dr. Romberg !!!!!!!!!!!!!!!!!!!!!!!!!! (← yeah, yeah, enough) Dr. Romberg insists: "My dear Dr. Hinrichs! It is perfectly clear that you can throw a tea bag further than a safe!"
Dr. Hinrichs's reply: "Come now, Dr. Romberg! We don't want to forget the base assumption of a v_0 presumed to be equal in both cases!!!"

[38]Some maintain: "Newton and Leibniz overtook the religions with the sciences" – and with this book, according to Dr. Romberg, the sciences have also been overtaken!

The Newtonian axiom renders statics as a special case: Since the acceleration here is a = 0, then Σ F = 0. Or we read the equation the other way around: If the sum of the forces is zero, then acceleration is also zero! But if the sum of the forces is not zero, then we can calculate the acceleration on the right side of the equation.

Strictly speaking, we have already successfully examined the change in motion caused by an acting force: During a free fall, only the weight G = mg acts upon a body. With the equations describing the motion in free fall, we ended up with $a(t) = \ddot{y}(t) = -g$ as the result of acceleration. But we have already derived this with some difficulty from the energy theorem.

We'll use the example of the cherry pit in free fall once again:

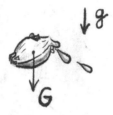

Figure 94: Free body diagram of the cherry pit

If you haven't deleted your entire knowledge of statics in the meantime with the usual C_2H_5OH-programs again like Dr. Romberg, then you can now immediately determine the acceleration of the cherry pit with the help of the new equation, in which the y-coordinate points up. So, sum up *all* of the forces in the direction of the coordinate (here y):

$$\Sigma F = -G = -mg = ma = m\,\ddot{y}, \qquad \Longrightarrow \quad \ddot{y} = -g \qquad .$$

Even if Dr. Hinrichs denies it: This is much simpler than in chapter 3.2.1.1! With these kinds of accelerated or delayed motions you first have to practice the art of drawing a free body diagram like in statics. A very basic procedure of mechanics is engaged: the method of drawing such a diagram. Without this procedure we can't go any further here, thus the drawing of a free body diagram is of special importance here.

The "force of acceleration" (m ÿ) is
n o t **entered into the free body diagram!**

So the next step is the creation of a free body diagram. Subsequently, we will apply the so-called Newtonian axiom.

And here's an example:

For some reason the pants pictured above have suddenly become heavier.[39, 40] The question is now, when the sliding pants will arrive at the bottom. Little tip: leg length: 1 m, mass of the pants: $m = 1$ kg, normal force between pants and legs by elastic: $F_N = 10$ N, average friction coefficient between pants and legs: $\mu = 0.25$ (additionally occurring sliding effects will be disregarded here). So: the free body diagram and the application of the Newtonian axiom according to the following **recipe** (this is how one should always proceed, in order to avoid the sign problems):

---

[39] Similarities of the depicted person with Dr. Romberg have been retouched by him on purpose during drawing.
[40] Here too, one must simply be generous and overlook Dr. Hinrichs's "humor".

First, the "acceleration term" (here mÿ) is written on the *left* side of the equal sign. On the *right* side the forces are summed up, and those that point in the direction of the coordinate are counted *positively*!

So, (y points up):

$$m\ddot{y} = \Sigma F = \quad -m\,g + \mu\,F_N = -1 \cdot 10 + 0.25 \cdot 10 \text{ kg m/s}^2$$

$$\Longrightarrow \quad \ddot{y} = -7.5 \text{ m/s}^2$$

$$\Longrightarrow \quad y = -0.5 \cdot 10 \cdot t^2 + 1 \text{ m} \qquad \text{with } y(t^*) = 0$$

$$\Longrightarrow \quad t^* = \sqrt{\tfrac{1}{5}} \text{ s} \quad = 0.4472 \text{ s}$$

So sometimes you can be caught with your pants down faster than you think!

183

Unfortunately, the turbo formula "Newtonian axiom", with which we've been able to accomplish a lot, only applies to translational motions – but what do you do when you want to describe a rotary motion?

Also in this case an expansion of the equations of statics applies in kinetics – we still remember from statics that the sum of moments must be zero. If this is not the case, we leave statics and a motion is brought into action: in this case a rotation. While for the translation the sum of the forces was proportional to acceleration (and still is, with the proportionality factor mass m), the sum of the factors is proportional to angular acceleration $\ddot{\varphi}$. The proportionality factor is in this case the moment of inertia J, which we already know (and in Dr. Hinrichs's case also love) from the observation of energy and can gather from the tables.

The so-called theorem of twist reads as follows – please memorize and never forget:

$$J^P \, \ddot{\varphi} = \Sigma \, M^P \qquad \text{(theorem of twist)}$$

Here some important tips from the old hand:

1) The same point of reference P has to apply to both sides of the equation!!! Careful with the signs! Always pin down the coordinates x, y, φ with directions first, then write the J$\ddot{\varphi}$-term on the left (analogous translation), and finally, on the right side, sum up the factors in this direction (of turn) with positive signs, and those going against this direction with negative signs. If you get a positive sum as a result, there will follow a circular acceleration in positive coordinate direction and vice versa!

2) This formula is to be enjoyed AT YOUR OWN RISK, since it is only applicable under certain conditions. The choice of point of reference for the theorem of twist is of great significance while using the formula (since the J and the sum of the factors is dependent on it). You can't go wrong, even without higher knowledge, if you:

☺) choose the body's center of gravity as point of reference,

☺☺) choose a point as point of reference for which the connecting line to the center of gravity is vertical to the acceleration of the center of gravity.

You choose between ☺) and ☺☺) according to the point upon which fewer unknown forces are acting.
One should always first check the velocity pole and the center of gravity as possible points!!!

We now recommend reading the previous lines at least 10 times, since that's half the battle – the rest is the accurate drawing of a free body diagram, beloved by all.

Here's an example to illustrate the point: the reserve beer keg rolling down into Dr. Romberg's cellar:

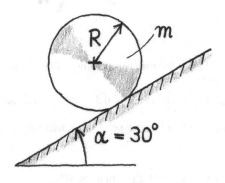

Figure 95: Rolling beer keg

Here's some more info: The beer keg can be understood as a rolling homogenous cylinder. The free body diagram represents the first difficulty here, which we will surely overcome.

But now to the choice of point of reference for our theorem of twist: According to rule 1) the center of gravity, i.e. the roll's center, can always be used; according to rule 2) the point of contact of the beer keg on the inclined plane could also be chosen (the center of gravity is accelerated in the direction of the inclined plane, which just this once is entered with a_S into the free body diagram. So the connecting line from the point of reference according to 2) to the center of gravity is vertical to the acceleration a_S). We'll begin by going against our rule, i.e. against our better judgment, with the center of gravity as point of reference:

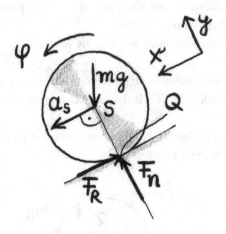

Figure 96: Free body diagram of the rolling beer keg

$$J^S \ddot{\varphi} = \Sigma M^S = F_R R \text{ with } J^S = 0.5 \text{ m R}^2 \qquad \text{(gathered from the table).}$$

Unfortunately F_R is still unknown, so Newton will help us again in determining the unknowns:

$$m \ddot{x} = \Sigma F_x = - F_R + m g \sin 30° \quad .$$

Unfortunately $\ddot{x}$ is still unknown, so we'll use kinematics:

$$\ddot{x} = R \ddot{\varphi} \quad .$$

Now we have to unite the three equations with the three unknowns F_R, $\ddot{x}$ and $\ddot{\varphi}$... We get the following result:

$$\ddot{\varphi} = \frac{g}{3R} \quad .$$

If, according to our recommendations, the choice of reference point according to either ☺) or ☹) is chosen according to where more unknown forces are acting, then the choice falls on the point of contact of the beer keg on the plane, since at this point two unknown forces F_N, F_R act. The point of contact is simultaneously the velocity pole of the beer keg – and is therefore denoted by Q.

Now the calculation mania begins: The theorem of twist reads in this case as follows:

$$J^Q \ddot{\phi} = \Sigma M^Q = m\, g\, R\, \sin 30°$$

with $J^Q =$ $0.5\, m\, R^2$ $+$ $m\, R^2$
       (gathered from table $+$ Steiner's share)

$$\Longrightarrow \quad \ddot{\phi} = \frac{g}{3R} \quad .$$

And that's it. By choosing the correct point of reference you can save yourself a lot of work. Whoever understands this can file away the horror topics theorem of linear momentum (Newton) and theorem of twist for good. Just a couple tricky geometries and free body diagrams (see the problems in chapter 4.3) can shock us now (Dr. Hinrichs would like to emphasize that this wouldn't in any way shock him in the least)!

Let's take this opportunity to rest a little following this difficult ascent through extremely difficult terrain and gaze in wonder upon the panorama that presents itself to us. The following overwhelming view of a wonderful analogy offers itself here:

translation	rotation
path x=s	angle ϕ
velocity $\dot{x} = v$	angular velocity $\dot{\phi} = \omega$
acceleration $\ddot{x} = a$	angular acceleration $\ddot{\phi} = \dot{\omega}$
mass m	moment of inertia J
force F	moment M
theorem of linear momentum $m\,\ddot{x} = \Sigma F_x$	theorem of twist $J^P \ddot{\phi} = \Sigma M^P$
kinetic energy $E_{trans} = \frac{1}{2}\, m\, \dot{x}^2$	kinetic energy $E_{rot} = \frac{1}{2}\, J\, \dot{\phi}^2$

After the loser has collected himself a bit, the expression "It's that easy..." spontaneously crosses his lips. But last but not least, we now come to a further topic: impact!

3.4 Impact

Impact is defined as the momentary collision of two bodies. During the very short impact duration Δt, very large forces are at work – other forces (e.g. weight) are negligible in comparison – and the position of the bodies involved in the impact does not change. [4]

A scientific discussion of the picture's content leads to the following result: The fist moves in the direction of the face and exerts forces onto it during contact. As a result of the contact forces the whole effort is directed onto parts of the face: Deformations appear, things are turned inside out, the teeth fall out,... If the face belongs to a portly gentleman, he'll probably stay standing where he was. But a pipsqueak will more or less be moved. He might even do a backwards somersault. You'll get some technical terms in a bit that you can impress with at the boxing ring: The backwards somersault takes place because the blow is not directed at the center of gravity. This is also referred to as an *eccentric impact.*[41] But if the blow penetrates deep into the stomach area, about where the center of gravity should lie, the professional speaks of a ~~STRIKE~~ *centric impact* (Dr. Hinrichs interjects that his own center of gravity lies elsewhere).

It will certainly be of interest to the mechanic to know how much of the blow is absorbed by the head and at what velocity the affected body then falls to the ground (backwards). The laws of impact provide the appropriate equations for this:

First, we have to alter the Newtonian axiom by integrating it (we aren't the first ones who've ever done it; we're copying again!):

$$\int_0^{t^*} F \, dt = \int_0^{t^*} m\ddot{x} \, dt$$

$$\int_0^{t^*} F \, dt = m\, \dot{x}(t{=}t^*) - m\, \dot{x}(t{=}0)$$

$$\int_0^{t^*} F \, dt = m\, v_2 - m\, v_1 \ .$$

You can interpret the whole thing like this: With this new formula (the so-called momentum theorem in integral form) we can determine the change in a quantity of motion through the acting of a force in a time interval of $t = 0$ to $t = t^*$. However, this quantity of motion is in a till now unusual form: m v. This quantity of motion is also referred to as *momentum.* So, with the momentum theorem described we can determine the *change in momentum* as a result of a force acting upon a body. The equation teaches us furthermore that without the exertion of a force (or for forces that are zero in the mean time!), the body's momentum must remain the same[42]:

$$m\, v_2 = m\, v_1.$$

---

[41] By the way: Dr. Romberg also has a slightly eccentric impact!
[42] Original sound byte of Dr. Romberg: "We have to detextbookify this part a bit more!" Dr. Hinrichs's remark: "I detextbookify – you detextbookify – he, she, it will have detextbookified..."

We can immediately apply this new knowledge to the face demolished by the punch. We draw a free body diagram of one of the involved bodies – either the hand or the face. The force progression on this body part will have the following characteristics over time (figure 97):

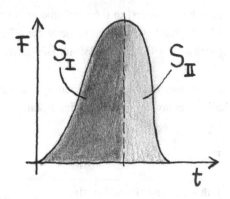

Figure 97: Characteristics over time of the force during momentum

For the elastic face and fist the force in the area has a symmetrical progression of contact. The fist goes into the face and bounces back out. On the way out the force of contact decreases. In the case of elasticity $S_I = S_{II}$ applies to the change in momentum (momentum = integral F over time t, i.e. the hatched area under the curve). If the face is *fully plastic*, as if it were made out of plasticine, then $S_{II} = 0$ and the fist's contour remains in the face after the blow. If the face is *partly plastic*, then $S_I > S_{II}$. In order to ascertain the "impact plasticity", the impact figure e is introduced:

$$S_{II} = e\, S_I .$$

Yup, and that means:

elastic impact: $e = 1$
partly plastic impact: $0 < e < 1$
fully plastic impact: $e = 0$

And what good is all this? Well, now, with the impact figure you can not only apply the momentum theorem in integral form to *one* body, but you can also illustrate the impact process between two bodies with it. In the textbooks in our bibliography wonderful ~~aggra~~ derivations are depicted that, for the case of the central impact of two bodies, lead to the following equations:

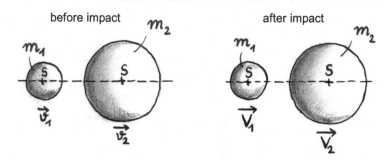

Figure 98: Impact of two elastic bodies

So the following concerns two bodies with masses m_1, m_2, that have the velocities v_1 and v_2 *before* and the velocities V_1 und V_2 *after* the impact. So now – please just believe it (Dr. Hinrichs murmurs: "Believing is not knowing", amateur theologian Dr. Romberg reassures: "Believing is trusting..."):

You can determine the impact figure e from the velocity differences:

$$e = -\frac{V_1 - V_2}{v_1 - v_2} \quad .$$

For the velocities after the impact the following applies:

$$V_1 = \frac{1}{m_1 + m_2}\left[(m_1 - e\,m_2)\,v_1 + (1 + e)\,m_2\,v_2\right] \quad ,$$

$$V_2 = \frac{1}{m_1 + m_2}\left[(1 + e)\,m_1\,v_1 + (m_2 - e\,m_1)\,v_2\right] \quad .$$

Energy loss during the impact:

$$\Delta T = \frac{1 - e^2}{2}\,\frac{m_1\,m_2}{m_1 + m_2}\,(v_1 - v_2)^2 \quad .$$

The simplified equations for the impact of body 1 against rigid wall 2 (borderline case $m_2 = \infty$, $v_2 = 0$) can save you as well:

$$V_1 = -e\,v_1, \qquad \Delta T = \frac{1-e^2}{2}\,m_1 v_1^2.$$

That was hard stuff! On the other hand, the application of these equations is relatively easy, since it follows the same old scheme!

Two quick examples:

In order to determine impact figure e, a ball is dropped from height H onto a plane. After the impact with the plane the ball reaches a maximum altitude h. How big is impact figure e ?

For this we of course need the velocities directly before and after the impact. They are:

$$mgH = \tfrac{1}{2}m\,v_1{}^2 \qquad \text{(energy theorem)}$$

$$\Longrightarrow \quad v_1 = \sqrt{2gH}, \text{ the same holds for } V_1 = \sqrt{2gh} \quad ,$$

from $V_1 = -\,e\,v_1$ \qquad follows $e = \sqrt{\dfrac{h}{H}}$.

And here's the next problem:

In the men's shower, a bar of soap (mass m, $v_m{=}0$) is lying on the slippery floor ($\mu{=}0$) at distance L from the wall. During a game of soap soccer a second bar of soap (mass M = 4 m) is kicked at the first bar at velocity v_M. Where will the bars of soap hit each other the *second* time, if the impacts between the bars are elastic (e = 1: hard soap – because it really does have to be harrrrrrd)?

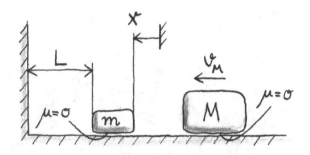

And now on to the soap equations: following the first impact:

$$V_M = \frac{1}{M+m}\big[(M-e\,m)\,v_M + (1+1)\,m\,0\big] = \frac{1}{5m}\big[3m\,v_M\big] = \frac{3}{5}\,v_M$$

$$V_m = \frac{1}{M+m}\big[(1+1)\,M\,v_M + (m-e\,M)\,0\big] = \frac{1}{5m}\big[8m\,v_M\big] = \frac{8}{5}\,v_M$$

Condition for the second impact: The traveled distances from M until the next impact at x_{impact} = the traveled distance from m to the wall and back to x_{impact}:

$$x_M\,(t_{impact}) = x_{impact}$$

$$x_m\,(t_{impact}) = L + L - x_{impact} = 2L - x_M$$

$$V_m\,t_{impact} = 2L - V_M\,t_{impact}$$

$$\frac{8}{5}\,v_M\;t_{impact} = 2L - \frac{3}{5}\,v_M\,t_{impact}$$

$$\Longrightarrow t_{impact} = \frac{10\,L}{11\,v_M}$$
$$\Longrightarrow x_{impact} = 6/11\,L .$$

Unbelievable ... that's all for the theory – or was it "feary"?

Since we know that the half-life of mechanics knowledge can sometimes be very short[43], please put on the sweatbands: It's time for some exercises!

---

4. Practice Makes the Loser a Winner

Clever education psychologists have discovered that the average loser retains on the average

> 10% through reading
> 20% through hearing
> 30% through seeing
> 50% through hearing and seeing
> and 90% through „falling flat on his/her face".

So now you can raise your degree of efficiency phenomenally by applying what you've read to the following exercises. However, in this case, the annoyance is already predetermined on your side.

„There is most likely nary a field in the engineering sciences in which one, after having ostensibly understood the theory, is as deceived and disappointed as in mechanics when it comes to solving practical problems" [23], see also the commentaries of all the losers during the announcement of the exam results.

But before you spontaneously burn this book after not being able to solve the first two exercises: The learning success sets in right at the point at which you can't get any further with an exercise (because if you had been able to solve the problem with no difficulty, you wouldn't have to calculate, since you've had it already. So you can always gleefully anticipate despairing over a problem. Completely savor the point of sheer desperation, try several solution paths, which may all lead into a morass and result in different answers. And then, after reading and dealing with the model solution, let 'er rip – that's what you call a learning success.[45]

---

[45] Of course you'll complain about the too-difficult exercises. Dr. Hinrichs let himself get carried away by his sadism while choosing the problems. Dr. Romberg thinks that though he himself has no desire for difficult problems, you can learn the most from difficult, practically unsolvable exercises.

You'll find a chapter number with every exercise. This tells you to which chapter (inclusive) you need to have read in order to make any sense to start working on the problem.

You'll also see a symbol before each exercise that gives the degree of difficulty of the problem, namely

☼: The exercises marked with this symbol are an absolute must! You should be able to at least partially solve these problems on your own following the reading of the indicated chapters – after studying the solution, the mistakes in your own answer should become clear.

💣*: These exercises are not to be underestimated – they make an impact (like a bomb), sometimes with destructive results.

☠: These exercises are a private source of joy for Dr. Hinrichs – but at the same time also very dangerous for the ego, such that one must be warned before attempting them. Following a most probably failed, but still important attempt at solving the problem on one's own, the outlined solutions should be studied, since they contain interesting mechanical tricks.

♱: This is how exercises are marked that shock the loser – even after reading the book – so much that they could kill him.[46]

📖: For the sake of completeness, we've added exercises containing fundamentals of mechanics that aren't described in chapters 1 to 3. Attempting to solve them on your own is pointless – but studying the solution as a first introduction seems to make more sense to us than studying the secondary literature, if you simply want to be able to keep up with the conversation.

For the problems that are guaranteed to crop up while you're solving the exercises, you should differentiate between problems with setting up the equations (meaning problems with the mechanics) and problems with solving the hard-won equations (these are problems with the math). The former are

---

[46] By the way: There are no such exercises!

desired and must, as previously mentioned, be overcome. The latter are pretty insignificant. Well, not really. But you don't get a handle on these problems by reading this or other mechanics books several times. So, if you come across a somewhat adventurous integral or four equations with four unknowns, then the mechanics of the exercise has been taken care of and you can be content!

And now, just a few little tricks that result in a big success, with which you can simplify mechanics life.

- In statics, the geometry and acting quantities of force are often given with angle φ (e.g. a block on an inclined plane at angle φ). Experience shows here that the sine often becomes a cosine in the solution. This can be avoided by not choosing an angle φ in the proximity of 45° in the sketch of the free-body diagram (for $\varphi=45°$, the opposite angles of an equilateral triangle can't be distinguished from one another), but rather a distinctly smaller (or larger) angle!

- When checking the achieved result, you should conduct a test of plausibility. Using the example of the sloped plane, it's always good to insert the extreme values ($\varphi=0$, $\varphi=90°$) once. Then, if for $\varphi=0$ the determined grade resistance $F_{gr} = G \cos\varphi$ corresponds to the weight, then something's not right!

- You can save yourself a lot of frustration when the results come out for the elementary test in Czech mechanics by running a unit check on the obtained result at the end of an exercise – and then slightly correct if need be.

- In a correct result, all of the quantities given in the text of the problem should be present. An innovative solution with new quantities is usually rewarded with a special deduction.

And now – knock 'em dead!

4.1 Statics Exercises

Dr. Romberg made a discovery, an „After-Five-Invention" (see Foreword, pp. VI): Instead of using a motor he hangs a magnet in front of his car to pull it, with the intent that the vehicle moves forwards.

Question: Can something like this work?

Solution

Of course not! Common sense (in some people more, in others less developed) already tells us that something's fishy here. Let's try it more formally: We could cut the car of the original question free from the magnet (cutting free always makes a good impression!)

Yes, there is a magnetic attraction at work on the vehicle. But it's devastating for the inventor that the magnet naturally has to prop itself up above the beam on the car.

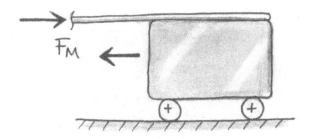

F_M

Unfortunately there's an opposing force to the magnetic attraction at work here. And the sum of the horizontal forces equals zero, so the car doesn't move. (Otherwise this exercise wouldn't be in the statics section and you'd have a perpetual motion machine). It would be more helpful to bind a donkey to the car and to hold a carrot in front of it just out of reach (but this isn't a perpetual motion machine either, since you have to provide food and water to the donkey)!

In order to save the invention, you could use an additional magnet.

And for constructive reasons we'll put both magnets in the car.

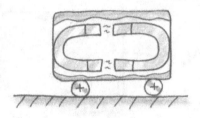

Now the question presents itself: If the invention is supposed to function, in which direction does the car move? (Dr. Hinrichs is already becoming impatient and has had his hand in the air the whole time waiting to be called on! Dr. Romberg responds to this gesture with his middle finger...) [cf. 30]

Exercise: 2   **Chapter: 1.4,   Degree of Difficulty: ✿**

Dr. Romberg, friend of animals and herbivore, offers a home to some vermin in an old bottle.

In the closed bottle, insects lie spread out on the bottom, while others are taking a flight around the bottle. For pseudoscientific purposes, the bottle is put on a scale (see also [30]).

Is the bottle

a)   heavier when all of the flies are on the bottom?
b)   heavier when all of the flies are flying around in the glass?
c)   heavier at the moment when all of the flies, startled by external noise and knocks, fly from the bottom into the air?

Solution

The weight remains the same regardless of whether the flies sit on the bottom or fly around. Because when the flies are in the air, we can make a nice free-body diagram for one of the flies:

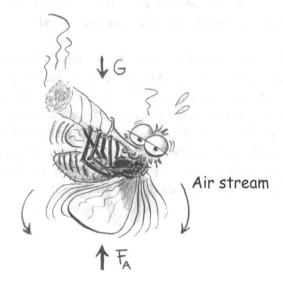

Air stream

In order for the fly to retain its state of suspension and not to hit the bottom, the fly's weight has to be counteracted by a force with a corresponding weight. And where does this come from? Out of thin air, so to speak! The air exerts a force on the fly (this is where the air friction on the fly's body, the fluid mechanics, the boost and such come into play). When the fly flaps with its wings, it sets an air stream in motion that is directed towards the bottom of the bottle and hits it. And in the end the air must exert exactly the same weight on the bottom of the bottle as on the fly. As mechanics we can regard bottles in statics as black boxes.

The condition for these considerations is, however, that the bottle is in a stationary state – this means that all flies remain approximately in the same flying altitude and the movement of the air is also relatively constant. But it's a different matter if we observe a nonstationary state (cf. c). This is now no longer a matter of statics – but common sense (or chapter 3) is of help here: Of course the flies push off from the bottom – at this moment a larger force is acting upon the scale. Stated differently: The bottle's center of gravity moves slightly upwards. And a change in the center of gravity can only be achieved by a force. This force must be exerted from the scale onto the bottle. What's important here is: In the mean time, the weight remains constant – since the flies sometimes fly against the cap or slow down beforehand.

In order to better understand the whole thing, you can also imagine a rotating propeller that's attached to a cap on the inside of a closed container. It makes no difference what degree of efficiency the propeller has, but the container will never fly into the air, even though a propulsive force is acting upwards on the cap. This is, however, annulled by the air masses blown downwards. So you'd have to remove the bottom of the container and perforate the cap (because of the air supply)... Then you'd have something like a simple engine that, structured in this manner, could actually fly (amazement!)

Dr. Romberg would like to lift a vat of apple juice[47] (mass M =100 kg) to the height of 1 m. In order not to overburden his circulation, he wants to roll the vat over an inclined plane of length L = 2m instead of lifting it.

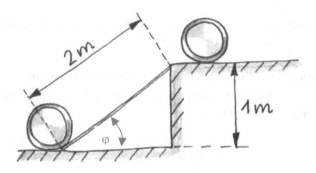

With what force does the vat have to be "rolled up" the plane? Additional question for chapter 3 experts: Determine the necessary force by way of the energy theorem!

| Solution |

First the easy way, i.e. without the energy theorem: You first have to apply at least the force which pulls the barrel down

$$F = Mg \sin \varphi,$$

---

[47] Dr. Hinrichs's remark: "That is completely absurd!"

where g ≈ 10 m/s² and

$$\sin \varphi = 1/2 \quad ,$$

so F = 500 N.

Alternatively, the energy theorem results in:

$$M\,g\,h = \int F ds = F \cdot 2m - F \cdot 0m \qquad ... => F = 500\ N \quad .$$

This exercise is pretty simple, but it demonstrates an interesting basic principle:

With an extension of the stretch from 1m (pure lifting) to 2m (inclined plane) we've attained a reduction of the necessary force. This basic principle is also the underlying principle of the pulley.

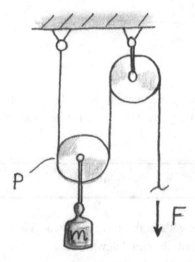

In cases involving a pulley, this bisection of force must also be compensated by a doubling of the stretch over which the rope has to be pulled. While statics with the equilibrium of forces seems to point to differing active principles (inclined plane: equilibrium of forces, pulley: free-body diagram for the pulley-wheel in motion, equilibrium of moments around the velocity pole P of the wheel in motion, try it out!), the above stated energy balance explains

$$E_{POT} = \int F ds$$

the circumstances for both cases:

In the case of the same potential energy to be attained, the doubling of the stretch goes along with a bisection of the force!

Exercise: 4   **Chapter: 1.4,   Degree of difficulty: ◆❋**

The homogenous cylinder W (weight G) is held in a state of rest on the inclined plane (angle of inclination β) by weight G via a weightless rope. What is the normal force between plane and cylinder?
Given: G, β=30°.

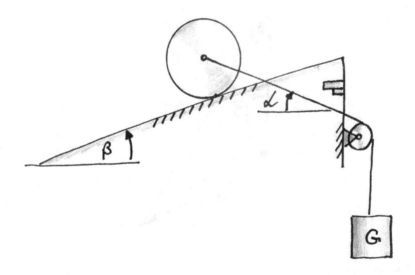

Free-body diagram:

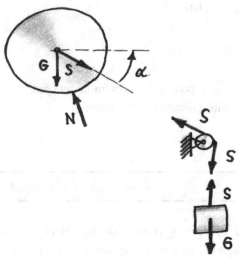

Equilibrium of forces on the weight: S = G

Force parallelogram for the cylinder:

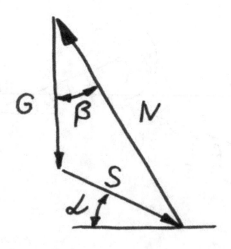

The rest is pure geometry:

$G \cos\alpha = N \sin\beta$

$G \sin\alpha = N \cos\beta - G$

$\cos\alpha = N \sin\beta / G = \dfrac{N}{2G}$

$\sin\alpha = N \cos\beta / G - 1 = \dfrac{N\sqrt{3}}{2G} - 1$

$\sin^2\alpha + \cos^2\alpha = 1 = \dfrac{N^2}{4G^2} + \dfrac{3N^2}{4G^2} - \dfrac{N\sqrt{3}}{G} + 1$

$\dfrac{N}{G}\left(\dfrac{N}{G} - \sqrt{3}\right) = 0$

Solutions: 1) $N = 0$, doesn't make sense!!!
             2) $N = \sqrt{3}G$

Exercise: 5   **Chapter: 1.4,   Degree of Difficulty:** 💣✳

The sketched friction-free system is burdened with force F.
Determine the reaction forces!
Given: a, F.

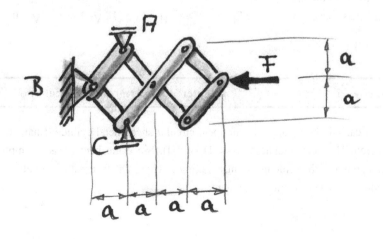

Free-body diagrams:

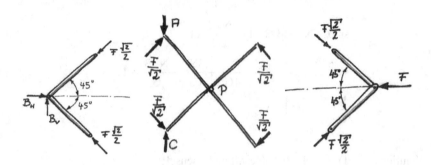

Note: The rods to the right and left are pendulum supports, and as such can only absorb forces lengthwise.

The sum of the horizontal forces for the free-body diagram to the right and the considerations of the symmetry (above = below) results in longitudinal force $F \sqrt{2}/2$.

Middle free-body diagram:

Symmetry or ΣM^P: $A = C$

ΣM^P for a pendulum support: $A = 2F$

Complete system: $\Sigma F_H = 0 \implies B_H = F$, $\Sigma F_V = 0 \implies B_V = 0$

Exercise: 6   **Chapter: 1.4,**   **Degree of Difficulty: ♠***

A beer can can be regarded as an open, circular sheet metal cylinder (diameter D, height H, sheet thickness s, s<<D, s<<H). Since the beer glass is empty once again and the replenishment is taking a while, Dr. Romberg combats his boredom with a scientific experiment:

The beer glass is unilaterally lifted on the lower edge. What is the maximum size that h_1 can attain without causing the container to fall over?
Given: G, D, H, s, s<<D, s<<H.

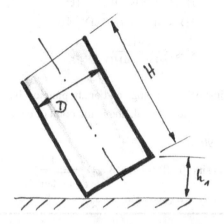

Center of gravity in the x-y system of coordinates:

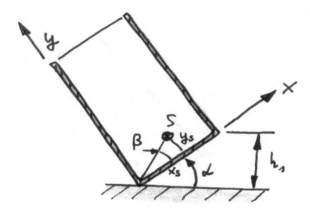

$$x_S = D/2 \qquad \text{(symmetry)}$$

$$y_S = \frac{\frac{s}{2}\frac{\pi}{4}(D-2s)^2 s + \frac{H}{2}\frac{\pi}{4}(D^2-(D-2s)^2)H}{\frac{\pi}{4}(D-2s)^2 s + \frac{\pi}{4}(D^2-(D-2s)^2)H}$$

$$= \frac{H/2 \quad \pi DHs + 0}{\pi D^2 s/4 + \pi DHs} = \frac{2H^2}{D+4H}$$

$$\tan\beta = y_S/x_S = \frac{4H^2}{D^2+4HD}$$

Unstable balance („just before tipping" and/or „tipping just beginning"): The center of gravity is direct, i.e. vertical, above the point of contact:

$$\alpha+\beta = 90°$$

$$h_1 = D \sin(90°-\beta) = D \cos\beta = D \frac{1}{\sqrt{1+\tan^2\beta}}$$

Exercise: 7   **Chapter: 1.4,**   **Degree of Difficulty: ◆***

In a depression (width a, depth a), a homogenous beam with a uniform cross-section of weight G rests as depicted. The system is frictionless.
What is the maximum length L the beam can possess for it not to slide out of the depression?
Given: a, G.

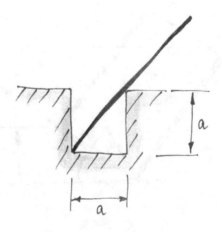

Free-body diagram:

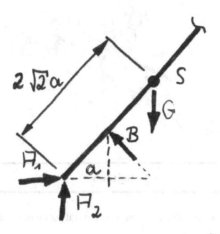

Beginning of the slide with $A_2 = 0$:   central force system

Procedure:

1) construction of the intersection A_1 - B

2) central system of forces for the center of gravity vertically above the intersection (only then is the sum of the factors around the intersection zero!)

$$\Longrightarrow L = 4\sqrt{2}\,a$$

A tractor (weight including driver G, center of gravity S) drives without the back drive wheels sliding at a constant speed up the slope (gradient angle α). In addition, the tractive force F is acting upon the tractor.

a) At which force F does the tractor tip over?

b) At least how big does the friction coefficient between the driving wheels and the plane have to be for the tractor not to slide before it tips over?

Given: a, G, α.

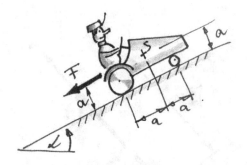

Solution

The tractor begins to tip over as soon as the axial force between front wheel and plane equals zero!

a) Free-body diagram:

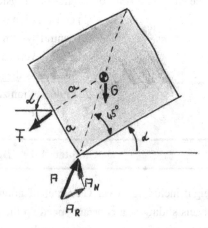

212

$$\Sigma \, M^A: \qquad - Fa + Ga \, (\cos\alpha - \sin\alpha) = 0$$

$$\implies \qquad F = G \, (\cos\alpha - \sin\alpha)$$

Alternative solution: a central system of forces:

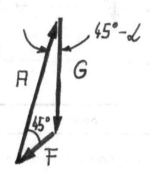

$$\implies \quad F = G \, \frac{\sin(45°-\alpha)}{\sin 45°} = \sqrt{2} \, G \sin(45°-\alpha)$$

(That's the same as in the first solution!!!)

b) $\qquad \mu_0 \geq \dfrac{A_R}{A_N} = \tan 45° = 1$

(μ_0 doesn't make any sense as far as the physics are concerned!)

Exercise: 9   **Chapter: 1.4,**   **Degree of Difficulty: ✸**

The depicted body (a thin homogenous triangular sheet with a massless rod) is to be held fast by force F in the diagrammed situation.
Determine the point of load incidence, the direction and the quantity of force F in the case of the smallest possible force. How large then is the resulting bearing reaction in bearing A?
Given: G, a.

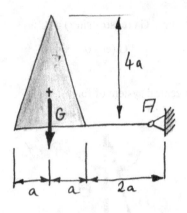

Solution

Consideration: - for minimization, the force must stand vertically on the lever arm

  - The lever arm must be at the maximum

From this, we deduce:

F has to act on the upper corner of the triangle and stand vertically on the connecting line between the apex and bearing A – in this case, the lever arm of force is at maximum!

Quantity of force: ΣM^A: $3\,G\,a - \sqrt{16a^2 + 9a^2}\ F = 0$

$\implies$ $F = \dfrac{3}{5}\,G$

Force parallelogram with forces F, A and G: $A = \dfrac{4}{5}\,G$

Exercise: 10 **Chapter: 1.8, Degree of Difficulty: ♠︎**

An ~~engineer noggin~~ hollow cube of uniform wall thickness (inner edge length a) hangs on a rope that is fastened to a corner of the cube. A sphere (weight G, diameter d<a) lies frictionless in its interior.

How big are the supporting forces acting on the sphere as regards quantity and slope in relation to the vertical?

Given: a, d, G.

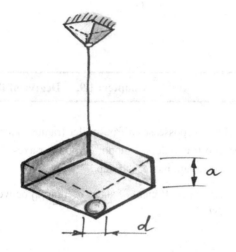

Solution

Approach 1:   System of coordinates in the direction of the cube edges:

$$\begin{bmatrix} N \\ 0 \\ 0 \end{bmatrix} + \begin{bmatrix} 0 \\ N \\ 0 \end{bmatrix} + \begin{bmatrix} 0 \\ 0 \\ N \end{bmatrix} = G \frac{1}{\sqrt{3}} \begin{bmatrix} 1 \\ 1 \\ 1 \end{bmatrix}$$

$$\implies \quad N = \frac{G}{\sqrt{3}}$$

Approach 2:   Angle φ between the cube diagonals and edge:

$$\cos\varphi = \frac{a}{\sqrt{a^2 + a^2 + a^2}} = \frac{1}{\sqrt{3}}$$

Each wall carries 1/3 of the weight:

$$N \cos\varphi = \frac{G}{3}$$

$$\Rightarrow \quad N = \frac{G}{\sqrt{3}}$$

Exercise: 11   **Chapter: 1.9,**   **Degree of Difficulty:** ◆⃰

A steamroller (weight G) possesses a front roller (radius r) and two rear drive rollers. The rollers are frictionlessly supported on their axes. The vehicle is to roll from the depicted situation over the step.

At least how big does the coefficient of static friction μ_0 between the road and the rollers have to be?

Given: L, r, G, α.

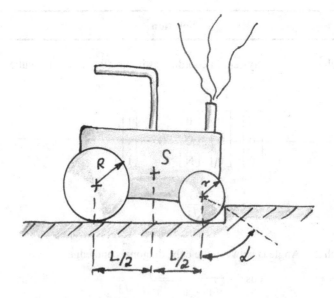

216

Free-body diagram:

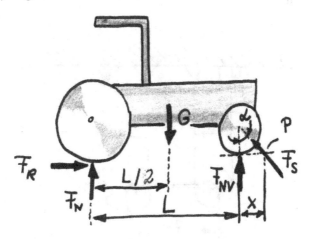

Beginning of the motion:	$F_{NV} = 0$
Normal force: ΣM^P:	$F_N (L+x) - G (L/2 + x) = 0$ with $x = r \sin \alpha$
	$F_N = G \dfrac{L/2 + r \sin \alpha}{L + r \sin \alpha}$
Friction force: ΣF_H, ΣF_V:	$F_R = F_{SV} \tan\alpha = (G - F_N) \tan\alpha$
	$= G \tan\alpha \dfrac{L/2}{L + r \sin \alpha}$
Coefficient of friction:	$\mu_0 \geq \dfrac{F_R}{F_N} = \dfrac{\tan \alpha}{1 + 2r \sin \alpha / L}$

Exercise: 12   **Chapter: 1.9,   Degree of Difficulty: ♠***

A body that is welded together from two disks (each of weight G, diameter D) and a weightless connecting rod is lowered to the ground with two equal winches W at a constant speed.

What is the bending moment in the middle of the connecting rod if the coefficient of friction μ acts between each disk and the corresponding rope?

Given: G, D, μ.

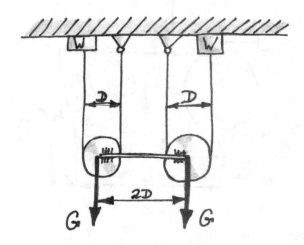

Free-body diagram left disk:

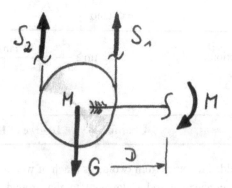

(From the free body diagram results that neither normal nor lateral force acts inside the beam connecting both disks!)

Rope friction:
$$\frac{S_1}{S_2} = e^{\mu\pi}$$

ΣF_V:
$$S_1 + S_2 = G$$

$$\implies S_2 = \frac{G}{1 + e^{\mu\pi}} \, , \, S_1 = G \frac{e^{\mu\pi}}{1 + e^{\mu\pi}}$$

ΣM^M:
$$M = \frac{D}{2}(S_1 - S_2) = \frac{GD}{2} \frac{e^{\mu\pi} - 1}{e^{\mu\pi} + 1}$$

Exercise: 13   **Chapter: 1.9,   Degree of Difficulty:** 🌢※

A belt drive (coefficient of friction μ) is pre-stressed with the depicted apparatus by weight G. What is the maximum output torque M_{AB} that can be transferred without causing the belt on one of the two pulleys to slide?
Given: d, D=4d, L=3d, μ, G.

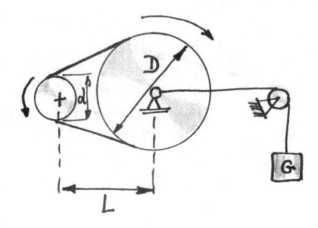

Geometry:

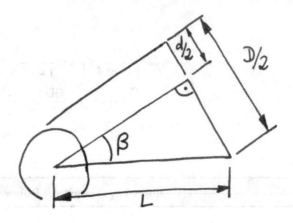

$$\sin\beta = \frac{D-d}{2L} = 0.5$$

$$\implies \quad \beta = 30° \text{ (corresponds to } \pi/6)$$

Angle of enlacement: $\alpha = \pi - 2\arcsin\beta = \frac{2}{3}\pi$

Free-body diagram driven pulley:

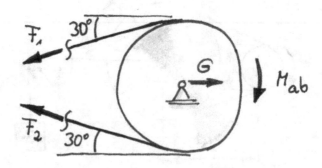

$$F_1 + F_2 = \frac{G}{\cos 30°} = \frac{2}{\sqrt{3}} G$$

$$F_1 - F_2 = \frac{2}{D} M_{ab}$$

$$F_{1max} = F_2 e^{\mu\alpha} = F_2 e^{\frac{2}{3}\mu\pi}$$

$$\Longrightarrow \qquad M_{AB} = \frac{GD}{\sqrt{3}} \frac{e^{\frac{2}{3}\mu\pi} - 1}{e^{\frac{2}{3}\mu\pi} + 1}$$

Exercise: 14   **Chapter: 1.9,   Degree of Difficulty:** 🌢☀

A frame (weight G) hangs as sketched by a thread on a nail.

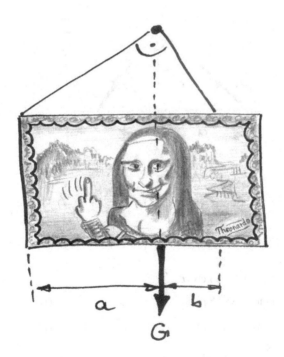

At least how big does the coefficient of friction μ between the nail and the thread have to be for the picture to hang horizontally if the suspension points on the frame are at different distances a and b from the picture's center of gravity?

Hint: The diameter of the nail can be disregarded for a and b, the same goes for the friction of the picture on the wall!

Given: a, b, a>b, G.

Solution

Free-body diagram, geometry:

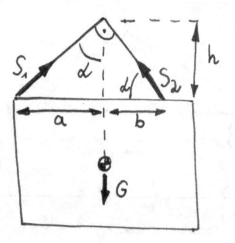

ΣF_H: $S_1 \sin\alpha = S_2 \cos\alpha$

   ==> $\tan\alpha = S_2 / S_1$ (I)

Geometry: $\tan\alpha = a/h = h/b$ ==> $\tan^2\alpha = a/b$ (II)

   I u. II: $\dfrac{S_2}{S_1} = \sqrt{\dfrac{a}{b}} > 1$

Rope friction: $\dfrac{S_2}{S_1} = e^{\mu\frac{\pi}{2}} = \sqrt{\dfrac{a}{b}}$

$\Longrightarrow \quad \mu = \dfrac{1}{\pi}\ln\dfrac{a}{b}$

Exercise: 15   **Chapter: 1.10,   Degree of Difficulty: 💣✳**

The depicted supporting structure in the margin consists of weightless rods and a homogenous triangular plate of uniform thickness and with weight G.

How large are the forces in the rods 1 to 3? Are we dealing here with tie rods or struts?

Given: a, G, F=G.

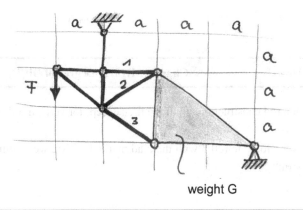

weight G

Solution

The sum of the moments around the right bearing (reaction force at upper left bearing is called A):

$$F\,4a - A\,3a + G\,\frac{2}{3}\,2a = 0$$

$$\Longrightarrow \quad A = \frac{16}{9}\,G$$

Ritter-cut:

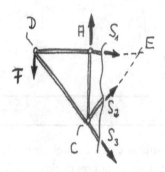

$$\Sigma M^C: \qquad S_1 = G \qquad\qquad\qquad \text{(tie rod)}$$

$$\Sigma M^D: \qquad S_2 = -\frac{1}{2}\sqrt{2}A = -\frac{8}{9}\sqrt{2}G \qquad \text{(strut)}$$

$$\Sigma M^E: \qquad \sqrt{2}\, a\, S_3 - A\, a + F\, 2a = 0$$

$$\qquad\qquad S_3 = -\frac{1}{9}\sqrt{2}\, G \qquad\qquad \text{(strut)}$$

Exercise: 16   **Chapter: 1.10,   Degree of Difficulty:** ✿

In the depicted rod construction, all of the rods except for vertical rod 5 possess length a.

How large is the force in rod 3 with the load marked F? Are we dealing with a tie rod or a strut?

Given: a, F.

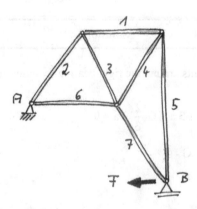

Support: $A_H = F$, $\Sigma M^A = 0 = Fa \sqrt{3}/2 - B a 3/2$
$$B = F / \sqrt{3} = - A_V$$

Ritter-cut:

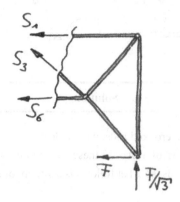

$$\Sigma F_V = S_3 \sqrt{3}/2 + F/\sqrt{3} = 0$$
$$S_3 = - 2F/3$$

Exercise: 17   **Chapter: 1.10,   Degree of Difficulty: ✿**

The rods 1 to 4 of the sketched system possess length r and are connected to one another by joint M. M is simultaneously the center of the sector, comprised by rods 5 to 8.

How large is the force in rod 9?

Given: F, r, angle see sketch.

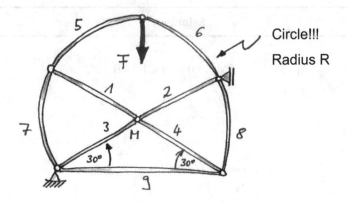

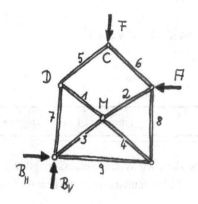

Trick: In the case of the crooked rods, we are dealing with pendulum supports. To clarify the directions of force of these pendulum supports, they will be replaced with straight rods, so what follows is a system of substitution:

Support: ΣF_V: $B_V \;=\; F$

ΣM^B: $A\,r - Fr\,\dfrac{\sqrt{3}}{2} \;=\; 0$

$\Longrightarrow$ $A = B_H = \dfrac{\sqrt{3}}{2}\,F$

Balance of junction points:

C: $S_5 = S_6 = -F$

D: $S_5 = S_7 = -F$

       $S_1 = -S_5 = F$

M: $S_4 = S_1 = F$

B: ΣF_V: $S_3 = 0$, $S_3 = S_2 \Rightarrow S_2 = 0$

       ΣF_H: $S_9 = -\dfrac{\sqrt{3}}{2} F$ (strut)

Exercise: 18   **Chapter: 1.10,   Degree of Difficulty: 🔥***

The depicted truss is acted upon by two single forces F. How large are the forces of rods 1, 2 and 3? Are we dealing here with tie rods, struts or zero-force rods?

Given: a, F.

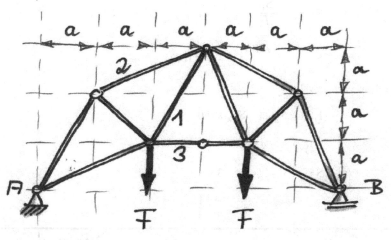

Solution

Support: $A_V = B = F$

Ritter-cut:

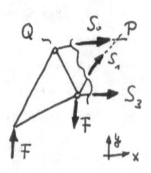

ΣM^P:

$3a\,F - a\,F - 2a\,S_3 = 0$

$S_3 \quad = F \qquad$ (tie rod)

ΣM^Q:

$-S_3\,a - S_{1x}\,a - S_{1y}\,a + F\,a + F\,a = 0$

$S_{1V} \quad = 2\,S_{1H}, \;\; S_{1H} = F/3, \;\; S_1{}^2 = S_{1H}{}^2 + S_{1V}{}^2 = 5\,S_{1H}{}^2$

$S_{1H} \quad = \dfrac{S_1}{\sqrt{5}}$

$\Longrightarrow \quad S_1 \quad = \dfrac{\sqrt{5}F}{3} \qquad$ (tie rod)

$\Sigma M^{\text{Point of load incidence}}$:

$F\,2a + S_{2y}\,a + S_{2x}\,a = 0$

with $\quad S_{2y} = S_2/\sqrt{5}$ and $S_{2x} = 2S_2/\sqrt{5}$

$\Longrightarrow \quad S_2 \quad = -2\,\sqrt{5}\,F/3$

Exercise: 19 **Chapter: 1.10,** **Degree of Difficulty:** 💣✷

The sketched flat truss bears a plate (weight G). How large are the rod forces 1 to 4? Are we dealing with tie rods or struts?

Given: a, G.

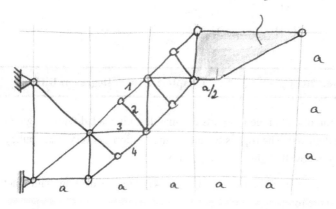

weight G

Solution

The disk's center of gravity:

$$x_S = \frac{\dfrac{a^2}{2}\dfrac{a}{4} + \dfrac{3a^2}{4}a}{\dfrac{a^2}{2} + \dfrac{3a^2}{4}} = 0.7\,a$$

Zero-force rod: $\qquad S_2 = 0$ $\qquad$ (reason: left joint of rod 2 cannot have a force in rod direction (where should the reaction force come from?))

Ritter-cut:

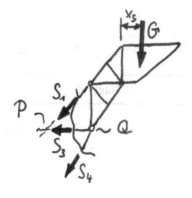

229

$$\Sigma M^Q: \qquad S_1 = 1.7 \sqrt{2}\ G$$
$$\Sigma M^P: \qquad S_4 = -2.7 \sqrt{2}\ G$$
$$\Sigma F_H: \qquad S_3 = G$$

Exercise: 20   **Chapter: 1.10,**   **Degree of Difficulty:** ✦✶

The pictured wall crane (all parts are massless) is loaded with weight G in the manner depicted. The rope is wound on the frictionless very small pulley R and hinged at the wall slightly above point A.

How large are the forces in rods 1 to 4? Are we dealing with traction or compressive forces?

Given: G, a, R.

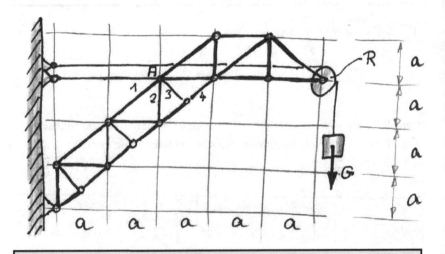

Solution

Zero-force rod: $S_3 = 0$ (right joint of rod 3 cannot have a force in rod direction)

230

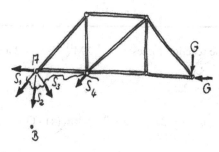

Bearing force A in horizontal rod (sum of moments around bearing left below):

$$A = \frac{2}{3}G \qquad \text{(traction)}$$

Ritter cut: ΣM^A :	$S_4 = -3\sqrt{2}\,G$	(strut)
ΣM^B :	$S_1 = \frac{4}{3}\sqrt{2}\,G$	(tie rod)
ΣF_V:	$S_2 = \frac{2}{3}G$	(tie rod)

Exercise: 21   **Chapter: 1.11,   Degree of Difficulty: ✹✶**

Two rigidly welded beams are acted upon by line load q_0 and force F.
Determine the bending moment $M_b(x)$ for the area $0 < x < 2a$ and sketch the course of moment for the horizontal beam.
Supply the largest occurring bending moment!
Given: a, F, q_0.

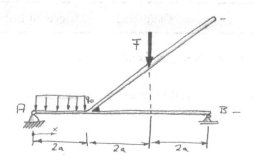

231

Determination of the reaction forces: $\Sigma M^B = F 2a + 2a\, q_0\, 5a - A_z\, 6a = 0$

$\Longrightarrow$ $A_z = F/3 + 5\, q_0\, a/3$

Lateral force for $0 < x < 2a$: $Q = A_z - q_0\, x$

Bending moment for $0 < x < 2a$: $\Sigma M^A = M_b(x) - Q\,x - \int_0^x \bar{x} q_0 d\bar{x} = 0$

$\Longrightarrow$ $M_b(x) = A_z\, x - q_0\, x^2 + 0.5\, q_0\, x^2$

$$= Fx/3 + 2q_0 a^2 \left[\frac{5x}{6a} - \left(\frac{x}{2a}\right)^2 \right]$$

Maximum bending moment: $M_{b,max} = M_b(2a) + F 2a = 8Fa/3 +$

$4q_0 a^2/3$

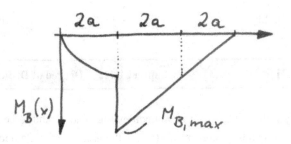

(Step in the course of moment, since in this place a factor is introduced by the beam protruding upwards at a slant.)

Exercise: 22   **Chapter: 1.11,**   **Degree of Difficulty:** ✦✶

A rotor (diameter d, D) of homogenous material with weight G is supported as shown.

Sketch the courses of the lateral force and the bending moment according to the dead weight and supply the values at the position a and in the middle of the rotor!

Given: a, d, D=2d, G.

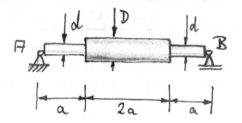

Weight per length:

$$4q\,2a + 2q\,a = G, \qquad \Longrightarrow \qquad q = \frac{G}{10a}$$

Line load:

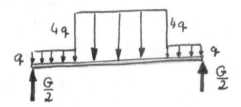

Course of lateral force:

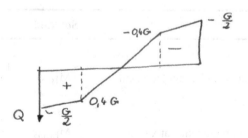

Course of bending moment:

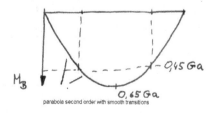

parabole second order with smooth transitions

233

It's five minutes to nine. Determine the course of the bending moment and its maximum value for the big hand of the church clock, assuming the hand (weight G) possesses uniform thickness and is triangular in shape.

Given: L, b, G.

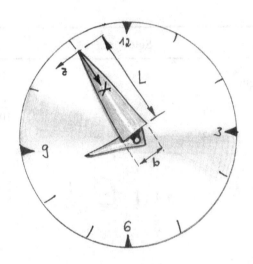

Solution

Course of bending moment:

$$M_B(x) = -\frac{1}{6}GL\left(\frac{x}{L}\right)^3$$

Maximum value at x = L:

$$M_{bmax} = -\,G\sin30°\,L/3 = -\frac{1}{6}GL$$

4.2 Elastostatics

A brake drum turns at constant angular velocity ω. The brake cable (cross-sectional area A) is stretched by means of the depicted mechanism.
What is the maximum possible weight G for the allowable tension σ_{allow} of the cable to not be exceeded?
Given: a, b, μ, A, σ_{allow}.

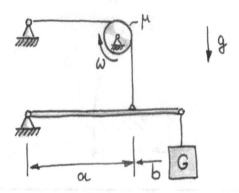

Solution

Force in the vertical section of the cable:
$$S_V = G \, \frac{a+b}{a}$$

Force in the horizontal section of the cable:
$$S_H = S_V \, e^{\,\mu\pi/2}$$

Tension: $\sigma_{allow} = \dfrac{S_H}{A} = \dfrac{G(a+b)e^{\mu\pi/2}}{aA}$

$\Longrightarrow \quad G = \dfrac{\sigma_{zul}\, a\, A\, e^{-\mu\pi/2}}{a+b}$

The compression stress σ_0 acts on the top side of the depicted truncated pyramid (edge length top side a, bottom side b, height h, E-module E). The dead weight of the truncated pyramid should be disregarded in the following.
a) How large is the tension σ_U on the bottom side?
b) By what amount is the stump shortened?
Given: a, b, h, σ_0, E.

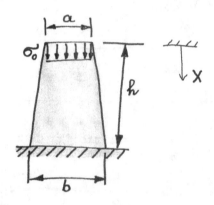

Solution

a)   Force on the top side:

$$F = \sigma_0\, a^2$$

Force on the bottom side:

$$F = \sigma_U\, b^2 \quad \Longrightarrow \quad \sigma_U = \sigma_0\, \frac{a^2}{b^2}$$

b) $$\Delta h = \int_0^h \frac{\sigma(x)}{E}\, dx \qquad \text{with} \quad \sigma(x) = \frac{F}{A(x)}$$

$$A(x) = s^2(x)$$

edge length $s(x) = a + x\,(b-a)\,/\,h$

$$\Delta h = \int\limits_0^h \frac{\sigma_0 a^2}{A(x)E}\, dx = \int\limits_0^h \frac{\sigma_0 a^2}{s^2(x)\,E}\, dx = \frac{\sigma_0 a^2}{E} \; \frac{h}{(a-b)} \; \frac{1}{a + x(b-a)/h}\Bigg|_0^h$$

(cf. Bronstein)

$$\Longrightarrow \qquad \Delta h = \frac{\sigma_0 a h}{E b} \qquad .$$

The stress-free mounted rod (length L, density ρ, cross-section A, E-module E, thermal expansion coefficient α) lies at point P on a base. How small is the friction coefficient μ at point P, if the rod, as a result of heating it by $\Delta\vartheta$, is elongated by ΔL.

Given: L, A, E, ρ, α, $\Delta\vartheta$, ΔL, g.

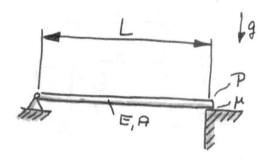

Solution

Here's the solution in short form:

Statics: friction force

$$F_R = \frac{\mu \rho A L g}{2}$$

Elongation:

$$\Delta L = -\frac{\mu \rho A L^2 g}{2EA} + \alpha \Delta \vartheta \, L$$

Solve for μ:

$$\mu = \frac{2E(\alpha \Delta \vartheta L - \Delta L)}{\rho L^2 g}$$

Exercise: 27   **Chapter: 2.2,   Degree of Difficulty:** ✿

The configuration shown consists of a rigid beam and two rods (E-module E, cross-sectional area A, thermal expansion coefficient α) that were mounted backlash- and stress-free at room temperature.

Which force acts in rod 1 if the ambient temperature is changed by $\Delta \vartheta$?

Given: E, A, α, $\Delta \vartheta$, a, b.

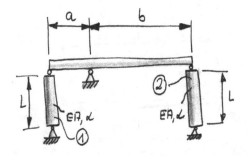

Solution

Normal forces in rods, balance of moments at the beam:

$$N_1 \, a = N_2 \, b$$

Geometry for change in angle of beam:

$$\Delta L_1 \, b = - \Delta L_2 \, a$$

238

Elongation of the rods:

$$\Delta L_1 \quad = \frac{N_1 L}{EA} + L\,\alpha\,\Delta\vartheta$$

$$\Delta L_2 \quad = \frac{N_2 L}{EA} + L\,\alpha\,\Delta\vartheta$$

i.e. four equations with the unknowns: $\quad N_1, N_2, \Delta L_1, \Delta L_2$

$$\Longrightarrow \qquad N_1 = -\,EA\alpha\Delta\vartheta\,\frac{(a+b)b}{a^2+b^2}\quad.$$

A conical rod with a circular cross-section lies backlash- and stress-free between two rigid walls. The rod is uniformly heated by $\Delta\vartheta$.
How large is the occurring maximum tension in the rod now?
Given: d, D, L, E, α, $\Delta\vartheta$, (L>>D).

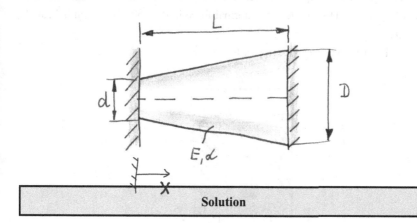

Solution

$$\Delta L \quad = \quad \int_0^L \left[\frac{N}{A(x)E} + \alpha\Delta\vartheta\right] dx \quad \text{with} \quad \Delta L \quad = 0,$$

$$\text{Cross section:} \qquad A(x) \quad = \pi\, r^2(x)$$

$$r(x) = \frac{1}{2}\left(d + \frac{D-d}{L}x\right)$$

$$\Longrightarrow \quad N = -\frac{\alpha\Delta\vartheta LE}{\int\frac{1}{A(x)}dx}$$

$$\text{with} \quad \int\frac{1}{A(x)}dx = \frac{L}{\pi dD}$$

(e.g. [Bronstein])

$$\Longrightarrow \quad N = -\alpha\Delta\vartheta E\pi dD$$

$$\Longrightarrow \quad \sigma_{max} = \frac{N}{A_{min}} = \frac{4N}{\pi d^2} = -4\,\alpha\,\Delta\vartheta\,E\,\frac{D}{d}\ .$$

Exercise: 29 **Chapter: 2.2,** **Degree of Difficulty:** ✦※

A screw joint consists of a socket (cross-sectional area A_H, E-module E_H, thermal expansion coefficient α_H) and a screw (cross-sectional area A_S, E-module E_S, thermal expansion coefficient α_S, pitch h). The nut is tightened by a sixth of a turn. How large is the normal force in the bolt if the joint is heated by $\Delta\vartheta$?

Given: E_S, E_H, A_S, A_H, α_S, α_H, h, L.

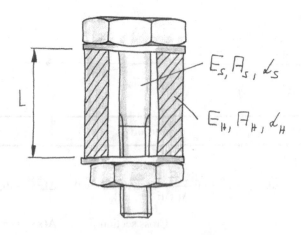

The difficulty with this exercise is (besides having to confront your innermost fears) neither statics nor the theory of strength of materials, but rather just the concept of what's actually going on here. A little trick: You'll comprehend this best if you keep the following special cases in mind:

1) That the socket is rigid: This would mean that the screw would have to be elongated by h/6 as a result of the bracing of the screw joint, i.e. $\Delta L_S = h/6$.

2) That the screw is rigid: This would mean that the socket would have to be shortened by h/6 as a result of the bracing of the screw joint, i.e. $\Delta L_H = h/6$.

So if both parts are elastic, then both deformations overlap one another:

$$\frac{h}{6} = \Delta L_S - \Delta L_H.$$

We can still determine the unknowns ΔL_S and ΔL_H:

$$\Delta L_S = \frac{NL}{E_S A_S} + \alpha_S \Delta\vartheta\, L,$$

$$\Delta L_H = \frac{-NL}{E_H A_H} + \alpha_H \Delta\vartheta\, L.$$

(Note: negative normal force in the socket, pressure load!)

Now we have to quickly run the three equations obtained through the mathematical mill, i.e. pack everything up and solve for N, get a common denominator.... This results in

$$N = \left[\frac{h}{6L} + (\alpha_H - \alpha_S)\,\Delta\vartheta\right]\; \frac{E_S A_S E_H A_H}{E_S A_S + E_H A_H}\;.$$

Finito!

The depicted component is stressed by the tensions σ_a, σ_b and τ as shown. The part has weld seams A and B that are inclined by α relative to the edge of the object. Determine the acting direct- and shear stresses σ_A, σ_B and τ_{AB} in the weld seams.

Given: σ_a = -2 kN/cm^2, σ_b = 6 kN/cm^2, τ = 3 kN/cm^2, α = 52°.

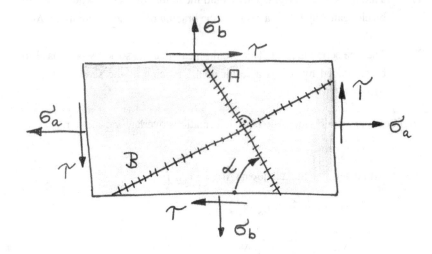

Solution

Construction of Mohr's circle of stress:

1) Plotting of point P_a (σ_a ; - τ) (note sign τ, cf. sign convention, chapter 2)

2) Plotting of point P_b (σ_b ; τ) (note sign τ, cf. sign convention, chapter 2)

3) P_a and P_b lie twisted at 90° relative to one another on the component ==> in Mohr's circle of stress, these points lie on opposite sides ==> the

242

intersection of the connecting line of P_a and P_b with the σ - axis forms the center of Mohr's circle of stress $\Longrightarrow$ draw the circle of stress around the center through P_a (and/ or P_b)

4) Plotting of P_B: From area b (this is the area where σ_b acts) you reach area A in the component through a clockwise turn by angle $\alpha=52°$ $\Longrightarrow$ in Mohr's circle of stress, we also have to turn clockwise, but by angle $2\alpha = 104°$ and get point

$$P_A \ (3.95 \ \text{kN/cm}^2; \ - 4.6 \ \text{N/cm}^2)$$

5) Point P_B lies opposite point P_A in Mohr's circle of stress ($2 \times 90° = 180°$). Stated differently: We come from area b to section B by turning counterclockwise by $90° - \alpha = 38°$, i.e. turning counterclockwise in Mohr's circle of stress by $2 \times 38° = 76°$ from P_b....

6) A reading of the coordinates leads to

$$P_B \ (0.1 \ \text{kN/cm}^2, \ 4.6 \ \text{kN/cm}^2)$$

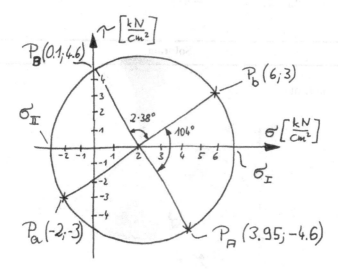

(Based on the illustration, exact values can also be calculated by way of the angular and circular relationships)

A thin sheet metal strip is strained as shown by an unknown tensile stress σ_0. In section A-A, turned by angle $\alpha = 22.5°$ relative to the unstrained edge, the direct stress $\sigma_n = 10.25$ N/mm² appears. In section B-B, turned by angle 3α relative to the unstrained edge, the same shear stress results as in A-A. What is the tension σ_0?

Given: $\sigma_n = 10.25$ N/mm².

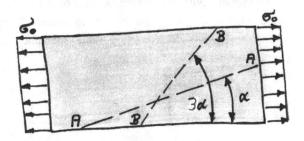

Solution

Mohr's circle of stress:

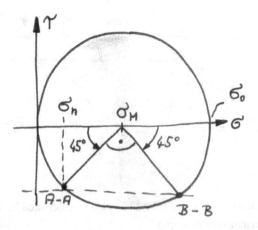

$$\Longrightarrow \qquad R = \sigma_M$$

$$\sigma_M = \sigma_n + R \cos 45°$$

$$\sigma_M = \frac{\sigma_n}{1 - 0.5\sqrt{2}}$$

$$\sigma_0 = 2\,\sigma_M = 2\,\frac{\sigma_n}{1 - 0.5\sqrt{2}} \approx 70 \text{ N/mm}^2$$

Exercise: 32   **Chapter: 2.3,**   **Degree of Difficulty:** ♠※

In a tensile test (rod cross-section A), the shear stresses in two intersecting planes inclined towards one another by angles 2α differ only in sign. How large is force F if the measured shear stresses amount to τ_m?

Given: τ_m, A, α.

Solution

$$\sin 2\alpha = \tau_m / R \Longrightarrow R = \tau_m / \sin 2\alpha$$

$$\sigma_I = 2R = F/A$$

$$F = 2AR = \frac{2A\tau_m}{\sin 2\alpha}$$

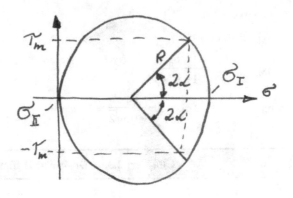

The tensions τ_a (σ_a=0) and σ_b (τ_b=0) act on the sketched tip of a thin plate. Determine from the given tension τ_a the direct stress σ_b and the equivalent stress according to the hypothesis of the largest modification of shape energy. Given: τ_a.

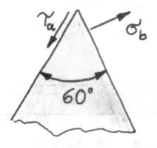

Solution

$$\sin 60° = \frac{\sqrt{3}}{2} = \tau_a / R$$

==>
$$R = \frac{2\tau_a}{\sqrt{3}} \quad , \quad \sigma_b = \sigma_I = R + R\cos 60° = \sqrt{3}\,\tau_a$$

$$\sigma_{II} = \sigma_I - 2R = -\frac{1}{\sqrt{3}}\,\tau_a$$

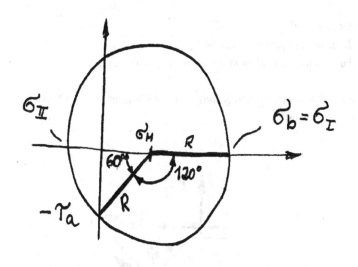

$$\sigma_V = \sqrt{0.5\left[\left(\sqrt{3}+\frac{1}{\sqrt{3}}\right)^2 + \left(\sqrt{3}\right)^2 + \left(\frac{1}{\sqrt{3}}\right)^2\right]}\,\tau_a$$

$$= \sqrt{\frac{13}{3}}\, \tau_a$$

The tensions σ_1, σ_2, σ_3, τ_1, τ_2 act on the sketched intersections of a body.

a) How large is angle φ?
b) How large are the principle stresses?
c) How large is the equivalent stress according to Tresca?

Given: $\sigma_1 = 60$ N/mm^2, $\sigma_2 = 10$ N/mm^2, $\sigma_3 = -85$ N/mm^2, $\tau_1 = 30$ N/mm^2,
    $\tau_2 = 20$ N/mm^2.

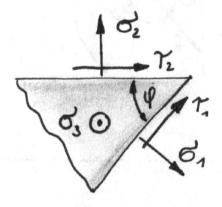

Well – we suggest a graphic solution here!!!!

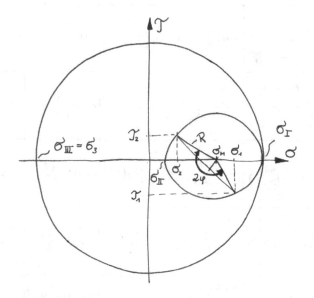

By means of reading off the drawing or through a mathematical solution, we get:

$$R = 10\sqrt{13} \text{ N/mm}^2,$$

$$\sigma_I = (4+\sqrt{13})\,10 \text{ N/mm}^2,$$

$$\sigma_{II} = (4-\sqrt{13})\,10 \text{ N/mm}^2,$$

$$\sigma_{III} = 85 \text{ N/mm}^2,$$

$$\varphi = 78{,}5°,$$

$$\sigma_{\text{Maximum shear stress criterion}} = \sigma_I - \sigma_{III} = 161 \text{ N/mm}^2 \ .$$

For the following exercises, the tables for the different bending cases, supplied in the literature, can be used. The following table is taken as an example from the formula collection of the Institute of Mechanics, University of Hannover, Germany [4]. The enumeration of the bending cases in the solutions of the exercises refers to the enumeration of the bending cases in the following table:

Load case	Equation of deflection line	Deflection	Slope
1	$w(x) = \frac{1}{6} \frac{F}{E \cdot I} \cdot \ell \cdot x^2 (3 - \frac{x}{\ell})$	$w(\ell) = f = \frac{F \cdot \ell^3}{3\,EI}$	$\tan \alpha = \frac{F \cdot \ell^2}{2\,EI}$
2	$w(x) = \frac{M}{2EI} \cdot x^2$	$w(\ell) = f = \frac{M \cdot \ell^2}{2\,EI}$	$\tan \alpha = \frac{M \cdot \ell}{E \cdot I}$
3	$w(x) = \frac{q \cdot \ell^4}{24EI} \left[6\left(\frac{x}{\ell}\right)^2 - 4\left(\frac{x}{\ell}\right)^3 + \left(\frac{x}{\ell}\right)^4 \right]$	$w(\ell) = f = \frac{q \cdot \ell^4}{8\,EI}$	$\tan \alpha = \frac{q \cdot \ell^3}{6\,EI}$
4	$x \leqq \ell/2$ $w(x) = \frac{F \cdot \ell^3}{16 \cdot EI} \left(\frac{x}{\ell} - \frac{4}{3} \cdot \frac{x^3}{\ell^3} \right)$	$w\left(\frac{\ell}{2}\right) = f = \frac{F \cdot \ell^3}{48\,EI}$	$\tan \alpha = \frac{F \cdot \ell^2}{16\,EI}$
5	$x \leqq a$: $w(x) = \frac{F \cdot \ell^3}{6EI} \cdot \frac{a}{\ell} \cdot \left(\frac{b}{\ell}\right)^2 \frac{x}{\ell} \left(1 + \frac{\ell}{b} - \frac{x^2}{ab}\right)$ $a \leqq x \leqq \ell$: $w(x) = \frac{F \ell^3}{6EI} \cdot \frac{b}{\ell} \cdot \left(\frac{a}{\ell}\right)^2 \frac{\ell-x}{\ell} \left(1 + \frac{\ell}{a} - \frac{(x-\ell)^2}{a \cdot b}\right)$	$w(a) = f = \frac{F \cdot \ell^3}{3EI}\left(\frac{a}{\ell}\right)^2 \left(\frac{b}{\ell}\right)^2$ $f^{*}_{max} = f \cdot \frac{\ell+b}{3b} \sqrt{\frac{\ell+b}{3a}}$ $x^{*}_{1max} = a \sqrt{\frac{(\ell+b)}{3a}}$ $*_{\underline{c}}$ holds for $a > b$	$\tan \alpha_1 = f \cdot \frac{1}{2b}\left(1 + \frac{\ell}{a}\right)$ $\tan \alpha_2 = f \cdot \frac{1}{2a}\left(1 + \frac{\ell}{b}\right)$
6	$w(x) = \frac{M_1 \cdot \ell^2}{6EI} \left[\frac{x}{\ell} - \frac{x^3}{\ell^3} \right] +$ $+ \frac{M_2 \cdot \ell^2}{6EI} \left[2\frac{x}{\ell} - 3\left(\frac{x}{\ell}\right)^2 + \left(\frac{x}{\ell}\right)^3 \right]$	for $M_1 = M_2$: $f_{max} = \frac{M_1 \cdot \ell^2}{8\,EI}$	$\tan \alpha_1 =$ $\frac{\ell}{6\,EI}(2M_1 + M_2)$ $\tan \alpha_2 =$ $\frac{\ell}{6\,EI}(2M_2 + M_1)$
7	$w(x) = \frac{q \cdot \ell^4}{24EI} \cdot \frac{x}{\ell} \left[1 - 2\left(\frac{x}{\ell}\right)^2 + \left(\frac{x}{\ell}\right)^3 \right]$	$w\left(\frac{\ell}{2}\right) = f_{max} = \frac{5 \cdot q \cdot \ell^4}{384\,EI}$	$\tan \alpha = \frac{q \cdot \ell^3}{24\,EI}$
8	$x \leqq \ell$: $w(x) = -\frac{F \cdot \ell^3}{6EI} \cdot \frac{a}{\ell} \cdot \frac{x}{\ell} \left[1 - \left(\frac{x}{\ell}\right)^2 \right]$ $\ell \leqq x \leqq (\ell + a)$: $w(x) = \frac{F \cdot \ell^3}{6EI} \cdot \frac{x-\ell}{\ell} \left[\frac{2a}{\ell} + \frac{3a}{\ell} \cdot \frac{x-\ell}{\ell} - \left(\frac{x-\ell}{\ell}\right)^2 \right]$	$f = \frac{F \cdot \ell^3}{3EI}\left(\frac{a}{\ell}\right)^2 \left(1 + \frac{a}{\ell}\right)$ $f_{max} = \frac{F \cdot \ell^3}{9\sqrt{3}\,EI} \cdot \frac{a}{\ell}$ $x_{1max} = \frac{\ell}{\sqrt{3}}$	$\tan \alpha_2 = \frac{F \cdot \ell^2}{6\,EI} \cdot \frac{a}{\ell}$ $\tan \alpha_1 = 2 \tan \alpha_2$

Two contact points (laminated springs with contact support) are arranged as shown. By what distance f would you have to vertically displace foot B of the right contact spring from the powerless situation depicted in order to produce the stipulated contact force F?

Given: F, L, EI.

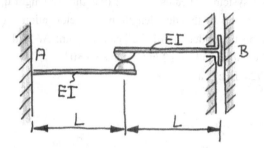

Solution

Loading case 1: Displacement of the left beam end: $f_{left} = f_{links} = \dfrac{FL^3}{3EI}$

Displacement of the right beam end: $f_{right} = f_{rechts} = \dfrac{FL^3}{3EI}$

Total displacement in B: $f_{TOT} = f_{left} + f_{right}$

$$\Longrightarrow \quad f_{TOT} = \dfrac{2FL^3}{3EI}$$

Alternative approach:

Substitute model of the cantilever beam:

Springs each with spring stiffness $c_{substitute} = \dfrac{3EI}{L^3}$

Connection of both springs: Spring deflections add up to
total displacement,
normal force equal in both springs

251

$$\frac{1}{c_{TOT}} = \frac{1}{c_{SUBSTITUTE}} + \frac{1}{c_{SUBSTITUTE}}, \dots c_{TOT} = \frac{3EI}{2L^3}$$

$$F = c_{TOT} \; f_{TOT} \Longrightarrow f_{TOT} = \frac{2FL^3}{3EI}$$

Exercise: 36 **Chapter: 2.5, Degree of Difficulty:** ✦❋

In the depicted system of beams 1 and 2 (weight per length q, E-module E, square cross-sections with edge length b) the elongation ε_A is measured longitudinally on the underside of the beam at point A. With which tractive force S must be pulled on beam 2 in C so that $\varepsilon_A=0$?

Given: F, q=F/2a, E, ε_A, a, b.

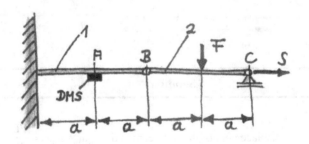

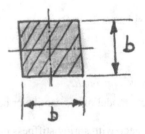

cross section of the beam

Solution

Statics: Bending moment in beam 1 at A:

$$M_{BA} = -\frac{5}{4} Fa$$

Tension on beam's underside:

$$\sigma_A = \frac{M_{BA}}{I} z + \frac{S}{A} = 0 \qquad \text{with} \quad A = b^2$$

$$I = b^4/12$$

$$z_{DMS} = b/2$$

$$\implies \quad S = \frac{15a}{2b} F$$

The depicted system is acted upon by force F. How large is the displacement of the point of load incidence?

Given: L, F, EI.

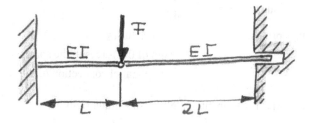

Solution

FKB joint ==> force in left and right beams, resolution of force:

$$F = F_{left} + F_{right}$$

Displacement left: loading case 1:

$$f_{left} = \frac{F_{links} L^3}{3EI}$$

253

Displacement right: loading case 1... !...:

$$f_{right} = \frac{F_{rechts}(2L)^3}{3EI}$$

Since the beam doesn't rip apart at the joint, you have $f = f_{left} = f_{right}$, i.e.

$$\frac{F_{links}L^3}{3EI} = \frac{F_{rechts}(2L)^3}{3EI},$$

$$\Longrightarrow \quad F_{left} = 8\,F_{right} = \frac{8}{9}\,F.$$

You can use this to determine the displacement:

$$f = \frac{8FL^3}{27EI}.$$

Alternative approach:

Substitute the cantilever beam with springs Determination of the rigidities of the substitute springs:

$$c_{substitute,\,left} = \frac{3EI}{L^3}$$

$$c_{substitute,\,right} = \frac{3EI}{8L^3}$$

Wiring of the springs: same paths,
sum of the forces in the springs
$\Longrightarrow$ parallel connection of the springs

$$c_{TOT} = c_{substitute,\,left} + c_{substitute,\,right} = ... = \frac{27EI}{8L^3}$$

$$F = c_{TOT}\,f$$

$$f = \frac{8FL^3}{27EI}$$

Exercise: 38	Chapter: 2.5, Degree of Difficulty: ✿

A laminated spring of sheet metal with uniform thickness d for a truck is to be assembled in a manner such that for the given load, the maximum tensions in each cross-section equal the allowable tension σ_{allow}.

Supply the width of the spring as a function of location.

Given: F, L, d, σ_{allow}.

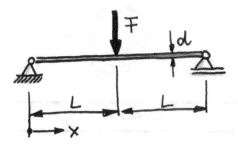

(For reasons of symmetry only the left side is taken into account here.)

Course of the bending moment for 0<x<L:

$$M_B(x) = \frac{F}{2}x$$

Maximum tension:

$$\sigma_{max} = \sigma_{allow} = \frac{M_B(x)}{I(x)} z_{max}$$

$$\text{with} \quad I(x) = \frac{b(x)d^3}{12},$$

$$z_{max} = d/2$$

$$\Longrightarrow \quad b(x) = \frac{3Fx}{\sigma_{zul}d^2}$$

Exercise: 39   **Chapter: 2.5,   Degree of Difficulty: 💣**

A laminated spring with uniform square cross-section (width B, thickness D, E-module E, length L>>D) is to be bent as shown into a circular ring.

a) Of which type does the load have to be?

b) How large is the maximum occurring tension?

Given: B, D, E, L>>D.

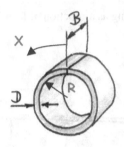

a) The radius of the bending line must be uniform, i.e. w''(x) = unif, i.e. $M_B(x)$ = unif. So a bending moment is triggered on the ends of the laminated spring.

$$w''(x) = 1/R = \frac{2\pi}{L} = -\frac{M_B}{EI} \implies M_B = (-)\frac{2\pi EI}{L}$$

b)
$$\sigma_{max} = \frac{M_B}{I} z_{max} \qquad \text{with } z_{max} = (-) D/2$$

$$\implies \quad \sigma_{max} = \frac{\pi ED}{L}$$

A glass tube (specific weight γ, inner diameter d, outer diameter D) is supported as shown.

What is the maximum length L the tube may have in order that the allowable tension σ_{allow} under a load is not transgressed by the dead weight?

Given: d, D, γ, σ_{allow}.

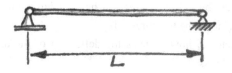

Tube cross section

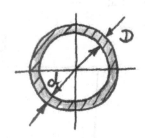

Solution

Section measures: maximum factor for x = L/2:

$$M_{max} = \frac{G}{2}\frac{L}{2} - \frac{G}{2}\frac{L}{4} = \frac{GL}{8} \quad \text{with} \quad G = \gamma\frac{1}{4}\left(D^2 - d^2\right)\pi L$$

Maximum bending stress:

$$\sigma_{max} = \frac{M_{max}}{I} z_{max} \quad \text{with} \quad z_{max} = D/2$$

$$\text{and} \quad I = \frac{\pi}{64}\left(D^4 - d^4\right)$$

$$\sigma_{max} = \frac{\gamma L^2 D}{(D^2 + d^2)} \leq \sigma_{allow}$$

$$\Longrightarrow \quad L \leq \sqrt{\frac{(D^2 + d^2)\sigma_{zul}}{\gamma \quad D}}$$

Exercise: 41   **Chapter: 2.5,   Degree of Difficulty: ✿◓☀**

The sketched wheel set is loaded with axle load F.

a) On which part of the axle does the largest tension occur?

b) At least how big does the axle's diameter d have to be in order for the allowable tension σ_{allow} not to be transgressed?

c) For the chosen diameter d, how large is the deflection of the axle in the middle between the wheel disks?

d) At which angle α do the wheel disks slant as a result of the load?

Given: F, s, L, σ_{allow}, E.

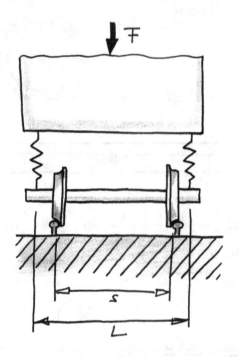

Solution

Hint: In the middle between the wheels we found bending case 6 of the table of bending cases with

$$M_1 = M_2 = \frac{F(s-L)}{4}$$

a) At the point of the maximum absolute value of bending moment, i.e. in the middle between the wheels. Upper edge: max. tensile stress, lower edge: max. compression stress.

b) Bending moment in the middle between the wheels:

$$M_{B,max} = \frac{F}{4}(s\text{-}L)$$

Tension as a result of bending:

$$\sigma = \frac{M_{B,max}}{I}z_{max} \le \sigma_{allow}$$

$$\implies \sigma_{allow} = \frac{4F(L-s)}{4\pi r^4}r$$

$$\implies d = 2r = 2\sqrt[3]{\frac{F(L-s)}{\pi\sigma_{zul}}} \quad .$$

c) cf. bending case 6:

$$f_{max} = \frac{F(s-L)s^2}{32EI}$$

$$= \frac{2F(s-L)s^2}{E\pi d^4}$$

d) cf. bending case 6:

$$\tan\alpha = \frac{F(s-L)s}{8EI}$$

The beam (bending rigidity EI, length 2L) is supported as shown by the rod (longitudinal rigidity EA, height h, thermal expansion coefficient α). At room temperature, the system is free of tension. The rod is heated by $\Delta\vartheta$. By what amount Δh does point P shift as a result of the heating of the rod?
Given: L, I, A, h, α, $\Delta\vartheta$.

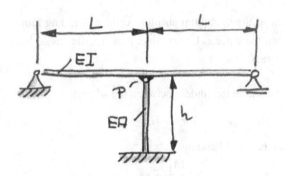

Cantilever beam: loading case 4:

$$\Delta h = \frac{F(2L)^3}{48EI} = \frac{FL^3}{6EI}$$

Rod:

$$\Delta h = -\frac{Fh}{EA} + h\,\alpha\,\Delta\vartheta$$

Introduction of F:

$$\Delta h = -\frac{6EIh}{EAL^3}\Delta h + h\,\alpha\,\Delta\vartheta$$

$$\Longrightarrow \quad \Delta h = \frac{h\alpha\Delta\vartheta}{1 + \dfrac{6Ih}{AL^3}}$$

Exercise: 43   **Chapter: 2.5,**   **Degree of Difficulty:** ✿

The beam (length L, side length a, wall thickness s) carries load F. How big is the maximum tension as a result of bending the beam?

Given: L=10m, F=200kN, s=10mm, a=300mm.

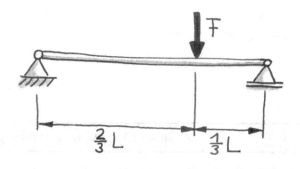

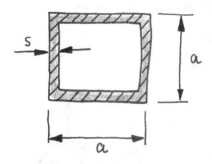

Solution

---

1) Determination of the maximum bending moment:

$$M_{max} = \frac{2}{9} \, FL \qquad \text{(location of the force transmission!)}$$

2) Moment of inertia of the box section:

$$I = \frac{a^4}{12} - \frac{(a - 2s)^4}{12}$$

3) Moment of resistance

$$W = \frac{I}{z_{max}} = \frac{2I}{a}$$

4) Maximum tension:

$$\sigma_{max} = \frac{M_{max}}{W}$$

5) Numerical value:

$$\sigma_{max} = 409{,}5 \ \text{N/mm}^2$$

Exercise: 44   **Chapter: 2.5,   Degree of Difficulty:** ✿♦✱

The diagram shows the simplified assembly of a force measurement device with strain gauges (DMS) 1 und 2. By means of the strain gauges, the difference of the extensions $\varepsilon_2 - \varepsilon_1$ is measured.

How large is force F?

Given: a, b, s, $\varepsilon_2 - \varepsilon_1$, E.

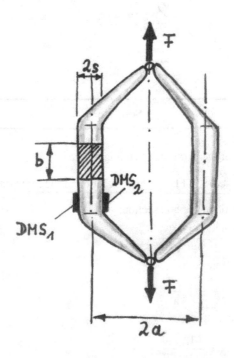

Substitute model:

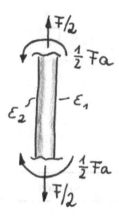

Tensions on the edge:

$$\sigma_{2,1} = \frac{N}{A} \pm \frac{M_B}{I} s \quad \text{with } I = 2bs^3 / 3$$

Difference of extension:

$$\varepsilon_2 - \varepsilon_1 = (\sigma_2 - \sigma_1) / E = \frac{2M_B s}{EI} = \frac{3Fa}{2Ebs^2}$$

Sought force:

$$F = \frac{2Ebs^2}{3a} (\varepsilon_2 - \varepsilon_1)$$

Exercise: 45   **Chapter: 2.5,**   **Degree of Difficulty:** 📖
------ **AN ABSOLUTE MUST!!! , BE SURE TO LOOK AT THIS !!!** ------

The exercises for determining the bending line that go beyond a simple reading from the table can be divided up into the following groups of exercises:

a) The overlapping of two bending cases on a component

The depicted beam (homogenous beam, mass m, bending rigidity EI, length L) is in addition to its weight acted upon by force F. How big does force F have to be in order for the dinting at the point of force transmission to disappear?
Given: m, EI, L.

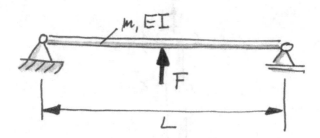

	Solution

Here, it's a matter of picking out the correct loading cases and browsing the right place in the table. The load due to concentrated force F represents bending case 4, so that

$$w_F(L/2) = -\frac{FL^3}{48EI}$$

is no big secret. Additionally, the beam's mass represents a line load of
$q = mg/L$. This is where we find bending case 7, which leads to

$$w_m(L/2) = \frac{5qL^4}{384EI},$$

the total dinting results from the overlapping of the individual cases, such that

$$w_{TOT} = -\frac{FL^3}{48EI} + \frac{5qL^4}{384EI} = 0 \quad.$$

You can determine the required force F from this equation:

$$F = \frac{5mg}{8}.$$

Too easy? O.K. Take this!

b) Overlapping of the bending cases on several beams

The depicted beam (bending rigidity EI, length of the sections each L) is acted upon by force F. How large is displacement u of the point of action in the direction of force F?

Given: F, L, EI.

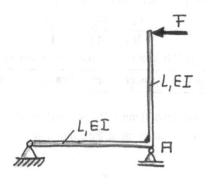

Solution

In this exercise, the loading cases are camouflaged somewhat better: First, we'll assume that *the horizontal section of the beam is rigid*. The beam is fixed in such a way on bearing A that it runs vertically upwards in any case. So an incline of the vertical section is avoided; a horizontal and vertical displacement of the beam at point A is avoided by the chosen mounting. If we regard the cutting reactions of the vertical section at point A, there is a lateral force and a bending moment here! To make a long story short: The load of the vertical section is represented by loading case 1. So, for the displacement of the loading point you get

$$u_I = \frac{FL^3}{3EI} \; .$$

In step two we'll assume that *the vertical section is rigid* while the horizontal beam is elastic. Now, however, a factor M=FL acts upon support A on the horizontal section that maltreats the beam according to loading case 6 such that

265

$$\tan\alpha_1 = \frac{FL^2}{3EI} \quad .$$

But the still-rigid vertical section then slants by angle α_1. As a result, the point of action shifts by

$$u_{II} = L \tan\alpha_1 = \frac{FL^3}{3EI} \quad ,$$

and the sought total displacement amounts to

$$u = u_I + u_{II} = \frac{FL^3}{3EI} + \frac{FL^3}{3EI} = \frac{2FL^3}{3EI} \quad .$$

So that was still too easy? In that case... no more Mr. Nice Guy!

) Statically overdetermined systems (special cases for a) and b)!)

The beam supported as shown (length 2L, bending rigidity EI) is acted upon by force F. Determine the displacement of the point of action.

Given: F, L, EI.

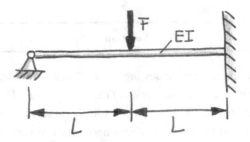

Solution

Well, now you're looking pretty clueless, huh? This loading case is unfortunately not included in our table. And then, to make matters worse, the fact that the system is statically undetermined is thrown into the bargain!!!

Can anyone solve an exercise like this at all? Not just anyone, but we can:

First we're going to play dumb (something that's not so difficult for Dr. Romberg) and ignore the floating bearing on the left side! Sure, here you get (loading case 1)

$$w_I(L) = \frac{FL^3}{3EI} \, ,$$

$$\tan\alpha = \frac{FL^2}{2EI}$$

$$w_I(2L) = w_I(L) + L\tan\alpha$$

$$= \frac{FL^3}{3EI} + \frac{FL^3}{2EI}$$

$$= \frac{5FL^3}{6EI}$$

And now for the climax: We simply substitute the floating bearing with force Q – whose size we don't know – yet. If we now forget about force F for a little bit, then the beam bends as a result of Q according to

$$w_{II}(L) = -\frac{5QL^3}{6EI} \quad ; \quad w_{II}(2L) = -\frac{Q(2L)^3}{3EI} \, .$$

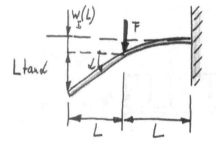
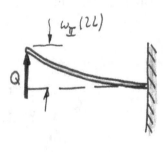

It's now imperative for $w_{TOT}(2L) = 0$ to remain intact, as long as the floating bearing serves its function. From

$$w_{TOT}(2L) = w_I(2L) + w_{II}(2L) = 0$$

you can determine the unknown bearing force Q and substitute the statically undetermined system with a statically determined system:

$$Q = \frac{5}{16} F \quad .$$

The dinting of the beam then results from the overlapping of both loading cases:

$$w_{TOT}(L) = w_I(L) + w_{II}(L) = \frac{7FL^3}{96EI} \quad .$$

(You could've gotten this in an easier way, cf. e.g. instruction manual).

The depicted beam (length L, bending rigidity EI) is loaded with a line load q_0 and a factor M. How large is the dinting at point $x = L/3$?
Given: q_0, M, L, EI.

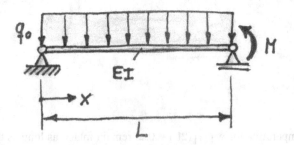

Overlapping (superposition) of load cases 6 und 7:

$$w_{TOT}(x) \quad = \quad w_6(x) + w_7(x)$$

$$\text{with} \quad w_6(x) \quad = \quad \frac{ML^2}{6EI}\left[\frac{x}{L} - \frac{x^3}{L^3}\right]$$

$$w_7(x) \quad = \frac{q_0 L^4}{24EI}\frac{x}{L}\left[1 - 2\left(\frac{x}{L}\right)^2 + \left(\frac{x}{L}\right)^3\right]$$

$$\Longrightarrow \quad w_{TOT}(x=L/3) = \quad \frac{11}{972}\frac{q_0 L^4}{EI} + \frac{4ML^2}{81EI}$$

Exercise: 47   **Chapter: 2.5,**   **Degree of Difficulty:** 🌑☀️⚲

A beam with uniform bending rigidity EI is loaded as shown with factor M.
How large is the bearing force in bearing B?
Given: a, M, EI.

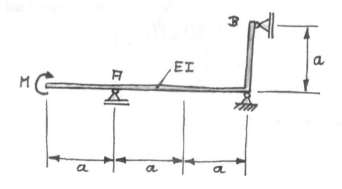

Overlapping (superposition!) of the partial loads:

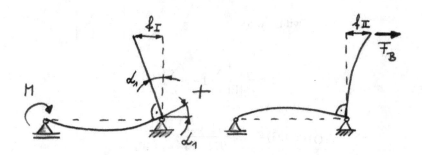

I: Loading case 6:

$$\tan\alpha_1 = \frac{2a}{6EI}M$$

$$f_I = \frac{Ma^2}{3EI}$$

II: Loading case 8:   (Surprise! And a bit of reflection...)
Agreed? If not, then choose the approach
with the combination of cases 1 und 6!)

$$f_{II} = \frac{F_B 8a^3}{3EI}\left(\frac{a}{2a}\right)^2\left(1+\frac{a}{2a}\right)$$

$$= \frac{F_B a^3}{EI}$$

No displacement of the bearing

$$\Longrightarrow \quad f_I = f_{II}$$

$$\Longrightarrow \quad F_B \quad = \quad \frac{M}{3a}$$

The depicted cantilever (E-module E, diameter d, length a) is acted upon at its free end with force F and factor M.
How big is the maximum deflection and the largest bending stress according to amount?

Given: a, d, E, F, M=2Fa/3.

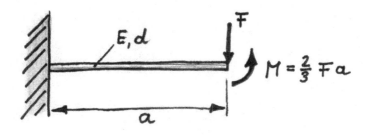

Solution

Overlapping (superposition) of bending cases 1 and 2:

$$w(x) = \frac{F}{6EI} ax^2 \left(3 - \frac{x}{a}\right) - \frac{M}{2EI} x^2 = \frac{Fax^2}{6EI} \left(1 - \frac{x}{a}\right)$$

$$\text{with } I = \frac{\pi d^4}{64}$$

Minimal deflection:

271

$$w'(x_{min}) = 0 = 2\,a\,x_{min} - 3\,x_{min}^2$$

$$\implies \quad x_{min} = \frac{2}{3}\,a$$

$$\implies \quad w_{max} = w(x_{min}) = \frac{128}{81\pi}\frac{Fa^3}{Ed^4}$$

The bending moment diminishes linearly in the direction of the bearing point $(M_B(x{=}0) = Fa/3)$.

Maximum bending moment:

$$M_{bmax} = \frac{2}{3}\,Fa$$

Bending stress:

$$\sigma_{bmax} = \frac{M_B}{I}\,z_{max} = \frac{64Fa}{3\pi d^3}$$

Given below is a supported beam jointed on both sides under a linear load. What is the equation of the bending line?

Given: q_1, q_2, EI, L.

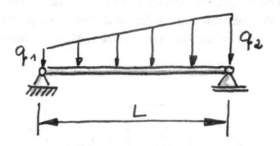

| Solution |

By the way: Looking it up in the instruction manual is not allowed here!

1) Line load:

$$q(x) = q_1 + (q_2 - q_1)\frac{x}{L}$$

2) Bearing reactions.

$$F_{right} = \frac{L}{6}q_1 + \frac{L}{3}q_2 \qquad , F_{left} = \frac{L}{3}q_1 + \frac{L}{6}q_2$$

3) Course of bending moment:

$$M_B(x) = F_{right}\, x - \left[\frac{q_1}{2}x^2 + \frac{q_2 - q_1}{3L}x^3 \right]$$

Yep, and then always integrate nicely... Boundary conditions: w(0)=0, w(L)=0

$$\implies \quad w(x) = \frac{1}{360}\frac{(q_2 - q_1)L^4}{EI}\left[3\frac{x^5}{L^5} - 10\frac{x^3}{L^3} + 7\frac{x}{L} \right]$$

$$+\frac{1}{24}\frac{q_1 L^4}{EI}\left[\frac{x^4}{L^4}-2\frac{x^3}{L^3}+\frac{x}{L}\right]$$

A supported beam jointed on both sides carries a load parabolically distributed over the length (peak value in the middle of the beam). Supply the equation of the bending line and calculate the dinting in the middle of the beam.

Given: q_0, L.

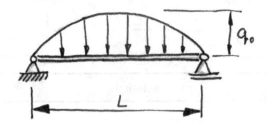

1) Line load:

$$q(x) \quad = Ax^2 + Bx + C,$$

Determination of A, B, C from $q(0)=q(L)=0$ and $q(L/2)=q_0$

$$q(x)= 4\, q_0 \left[\frac{x}{L}-\frac{x^2}{L^2}\right]$$

2) Support:

$$F_V \quad = \frac{1}{3}q_0 L$$

3) Bending moment:

274

$$M_B(x) = F_V x + \int_0^x q(x)x\,dx$$

$$= F_V x + 4\,q_0 \left[\frac{x^2}{2L} - \frac{x^3}{3L^2} \right]$$

Yep, and then integrate nicely again, boundary conditions $w(0) = w(L) = 0$

..... ==> $$w(L/2) = \frac{61}{5760}\,\frac{q_0 L^4}{EI}$$

Exercise: 51   **Chapter: 2.5,   Degree of Difficulty:** ◢※

A beam (length 3L, bending rigidity EI) is supported at A, B and C.
How large is the reaction in B if the beam is acted upon by force F as shown?
Given: L, EI, F.

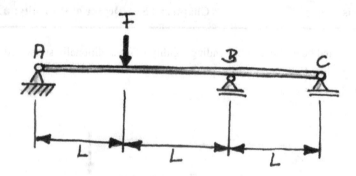

Solution

System statically overdetermined, so bearing B is substituted with vertical force B.

Superposition: I)   system without B with force F at x = L

275

II)   system without F with force B at x = 2L

These are both bending case 5, B has to be fiddled with in the sum of the individual bendings in such a manner that for the dinting at the place of the bearing $w_{TOT}(x=2L)=0$:

$$w_I(x=2L) = \frac{F(3L)^3}{6EI} \frac{2L}{3L}\left(\frac{L}{3L}\right)^2 \frac{3L-2L}{3L}\left(1+\frac{3L}{L}-\frac{(2L-3L)^2}{L\,2L}\right)$$

$$= \frac{7}{18}\frac{FL^3}{EI}$$

$$w_{II}(x=2L) = -\frac{F(3L)^3}{3EI}\left(\frac{2L}{3L}\right)^2\left(\frac{L}{3L}\right)^2 = -\frac{4BL^3}{9EI}$$

$$w_I(x=2L) + w_{II}(x=2L) = 0$$

$$\Longrightarrow B = \frac{7}{8}F$$

Exercise: 52   **Chapter: 2.5,   Degree of Difficulty:** ♦※

The depicted beam (length L, bending rigidity EI) is additionally supported by a spring. How large is the spring power?
Given: F, a, $c = EI/a^3$.

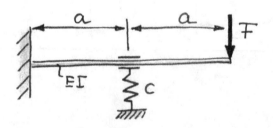

Solution

Deformation only by F without spring:

276

$$w_F(a) = \frac{1}{6}\frac{F}{EI}\,2a\,a^2\left(3-\frac{a}{2a}\right) = \frac{5}{6}\frac{Fa^3}{EI}$$

Deformation only by force F_F acting at the place of the spring:

$$w_c(a) = -\frac{1}{3}\frac{F_F a^3}{EI}\quad.$$

If the bending line of the total system is described by w_{TOT}, then the spring power is

$$F_F = c\, w_{TOT},$$

so
$$w_c(a) = -\frac{1}{3}\frac{c w_{TOT} a^3}{EI} = -\frac{1}{3} w_{TOT}$$

Overlapping of the load by the spring and force F:

$$w_{TOT}(a) = w_F(a) + w_c(a) = \frac{5}{6}\frac{Fa^3}{EI} - \frac{1}{3}w\,{TOT}$$

$$\implies\quad w_{TOT}(a) = \frac{15}{24}\frac{Fa^3}{EI}$$

$$\implies\quad F_F = c\, w_{TOT}(a) = \frac{5}{8}F\,.$$

Exercise: 53   **Chapter: 2.5,   Degree of Difficulty: 📖**

The depicted square beam (length L, bending rigidity EI, width B, height H), tightly clamped at angle α, is acted upon on its free end by force F. Determine the horizontal displacement of the point of load incidence.

Given: L, H, B, α, E.

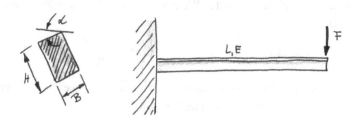

Let's do first what we've always done: We'll get the geometrical moments of inertia for bending around the axes of symmetry y and z from table 2:

$$I_y = BH^3/12 \quad , \qquad\qquad I_z = HB^3/12 \quad .$$

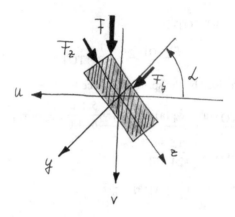

In statics, we combined two forces into one, the resulting force. Now we're going to do the opposite: We'll divide force F into two force components F_y and F_z:

$$F_y = F \sin\alpha \quad , \qquad\qquad F_z = F \cos\alpha \quad .$$

Since it doesn't matter to the beam whether it's acted upon by resulting force F or by the two force components, we have now with this little trick transformed the bending around any given axis in such a manner that we get two bendings around the principle axes y and z. We can then work this out with the standard load cases:

$$w_y(L) = \frac{F_y L^3}{3EI_y}, \qquad\qquad w_z(L) = \frac{F_z L^3}{3EI_z} \quad .$$

Now we can throw the equations together and recalculate the displacements in y and z-direction into the displacement in u-direction:

278

$$w_u(L) = w_y(L) \cos\alpha - w_z(L) \sin\alpha$$

$$= \frac{4FL^3 \sin\alpha \cos\alpha}{EBH}\left[\frac{1}{B^2} - \frac{1}{H^2}\right]$$

$$= \frac{2FL^3 \sin 2\alpha}{EBH}\left[\frac{1}{B^2} - \frac{1}{H^2}\right] \quad .$$

(Complementary to the vertical displacement:

$$w_v(L) = w_y(L) \sin\alpha + w_z(L) \cos\alpha$$

$$= \frac{4FL^3}{EBH}\left[\frac{\sin^2\alpha}{B^2} + \frac{\cos^2\alpha}{H^2}\right] \quad . \;)$$

(Interesting here is the control of:

        1)   straight bending, i.e. $\alpha = 0°$ or $\alpha = 90°$

        2)   straight bending for any given α and B=H,

              i.e. point symmetric bodies, $\sin^2\alpha + \cos^2\alpha = 1$)

So, through the utilization of statics, the skewed bending is strictly speaking nothing new.

The sketched profile is acted upon by force F. How large is distance d of force F from the center of gravity of the profile's surface if the profile, as a result of shear stress, is not twisted by the lateral force?

Given: a, t, t<<a.

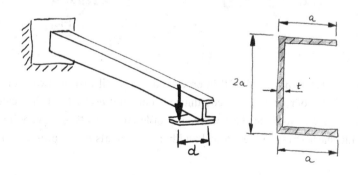

First, we'll get points by doing what we already can: calculating the geometrical moment of inertia:

$$I = \frac{t(2a)^3}{12} + 2\left(a^2\, a\, t\right) + \frac{at^3}{12}\,.$$

In the following, we'll leave out the last term, since for $t \ll a$ it can be disregarded relative to the other addends:

$$I = \frac{8ta^3}{3}$$

The resulting shear stress distribution looks like this:

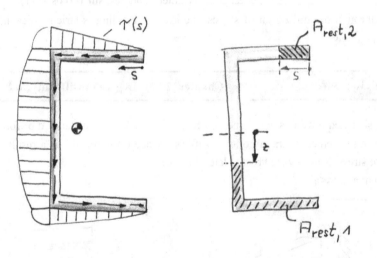

This now has to be calculated! In order to do so, we'll first put two coordinate systems into our profile: the coordinate z counting down from the center of gravity, as well as – for the horizontal profile components – the coordinate s. For the shear stress in the vertical and horizontal parts of the profile you now get

$$\tau(z) = -\frac{QS(z)}{Ib(z)} \qquad \text{resp.} \qquad \tau(s) = -\frac{QS(s)}{Ib(s)} \quad .$$

With the known geometrical moment of inertia, $b(z) = b(s) = t$ and $Q = F$ we're only missing the static factor. This results from the following:

vertical part:

$A_{rest,1} = t\,a + (a - z)\,t$

$z_{rest,1} = \dfrac{t\,a\,a + (a - z)\,t\,(z + (a - z)/2)}{A_{rest,1}}$

$S(z) = t\,a\,a + (a - z)\,t\,(z + (a-z)/2)$

$\qquad = 0.5\,t\,(3a^2 - z^2)$

$\tau(z) = -\dfrac{3Q(3 - z^2/a^2)}{16ta}$

horizontal webs:

$A_{rest,2} \qquad = t\,s$

$z_{rest,2} = a$

$S(s) = a\,t\,s$

$\tau(s) = -\dfrac{3Qats}{8\,t^2\,a^3}$

Now we can calculate the acting forces from the shear stress:
It follows for forces F_1, F_2 in the horizontal web of the profile

$$F_i \quad = \quad \int \tau dA = \int \tau\,t\,ds = \dots = \frac{3F}{16} \qquad . \ (i=1,2)$$

The direction of these forces corresponds with the directions of the shear stresses so that the horizontal forces are canceled out.
The resulting force F_3 of the thrusts in the vertical part of the beams results analogously:

$$F_3 \quad = \quad \int \tau dA = \int \tau\,t\,dz = \dots = F$$

So the resulting force of the three concentrated loads yields the lateral force!!!
The forces give rise to a factor around the center of area $y_s = 0.25\ a$:

$$M \quad = \quad \frac{3F}{16}a + \frac{3F}{16}a + \frac{F}{4}a = \frac{5Fa}{8}$$

This factor has to be balanced out with the force acting at distance d from the center of gravity:

$$F\,d \quad = \quad M \quad \Longrightarrow \quad d \quad = \quad \frac{5}{8}a.$$

281

By the way, the point of load incidence calculated in this way is also called shear center.

AND NOW FOR EVERYONE (EXCEPT DR. HINRICHS): Actually, you can forget the whole calculation!!! You should just take note of the fact that there exists a phenomenon like shear center.

Exercise: 55	Chapter: 2.6, Degree of Difficulty: ✿

For the depicted boiler the wall thickness s should be dimensioned in such a way that in the case of excess pressure Δp, the larger of the principle stresses does not transgress the highest permitted value σ_{allow}. How large is the maximum shear stress?

Given: Δp, σ_{allow}, d

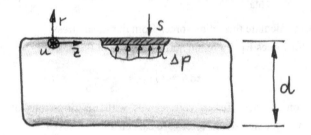

Solution

Tension in z-axial direction:

$$\sigma_z = \frac{\Delta p d}{4s}$$

Tension in circumferential direction:

$$\sigma_u = \frac{\Delta p d}{2s} > \sigma_z$$

$$\implies \quad \sigma_u \leq \sigma_{allow}$$

$$\implies \quad s \ge \frac{\Delta pd}{2\sigma_{zul}}$$

$$\tau = \sigma_u/4$$

A twisting moment M_t acts upon axle 1 of the sketched gear. Axle 1 (diameter d_1, shear modulus G) with a gear wheel (number of teeth z_1) rests above a further gear (number of teeth z_2) and is connected to the tightly clamped axle 2 (diameter d_2, shear modulus G). What is the contorsion φ of axle 1 at the point at which M_t is introduced into the axle?

Given: M_t, G, L_1, L_2, z_1, z_2, d_1, d_2.

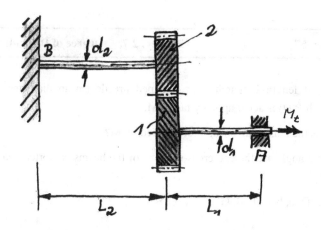

Solution

Twisting moment axle 2:

$$M_{t2} = M_t \frac{z_2}{z_1}$$

Contorsion axle 2:

$$\Delta\varphi_2 = \frac{M_{t2}L_2}{G\pi d_2^4 / 32}$$

Contorsion gear wheel 1:

$$\Delta\varphi_1 = \Delta\varphi_2 \frac{z_2}{z_1}$$

Total contorsion:

$$\Delta\varphi_{TOT} = \Delta\varphi_1 + \frac{M_t L_1}{G\pi d_1^4 / 32}$$

$$= \frac{32 M_t}{\pi G}\left[\frac{z_2^2}{z_1^2}\frac{L_2}{d_2^4} + \frac{L_1}{d_1^4}\right]$$

Exercise: 57   **Chapter: 2.7,   Degree of Difficulty: 🌶✳**

A beam of length L that has an e-shaped profile (mean diameter D, wall thickness b<<D) is acted upon by factor M.

a) What is the maximum shear stress caused by M?

b) At what angle are the end cross-sections of the beams contorted against one another?

Given: L, D, b, b<<D, M, G.

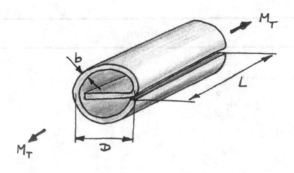

Torsion of a thin-walled slotted hollow section,

effective length of cross-section:

$$s = \pi D + D = (\pi + 1) D$$

Geometrical moment of inertia of torsion:

$$I_t = \frac{(\pi + 1)}{3} Db^3$$

a) Maximum shear stress:

$$\tau_{max} = \frac{M_T}{I_t} b = \frac{3}{\pi + 1} \frac{M}{Db^2}$$

b) Contorsion:

$$\Delta\varphi \quad = \quad \frac{M_T L}{GI_t}$$

$$= \quad \frac{3}{\pi + 1} \frac{ML}{GDb^3}$$

Exercise: 58   **Chapter: 2.7,   Degree of Difficulty: ✿💣✳**

A road sign mounted on a vertical post (thin-walled tube, outer diameter D) with the help of a frame is acted upon by wind power F as shown.

At least which wall thickness s must the post possess in order that – at the point of the maximum strain – the equivalent stress according to the hypothesis of the modification of shape energy doesn't transgress the permitted tension σ_{allow}?

Hint: Dead weight and shear stresses as a result of the lateral force are to be disregarded.

Given: a, h, D, F, σ_{allow}.

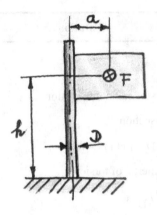

Solution

Maximum bending moment acts at the foot!

$$M_t = Fa \qquad\qquad M_{B,max} = F\,h$$

$$\tau_{max} = \frac{Fa}{I_t}\frac{D}{2} = 2a\,\frac{F}{\pi D^2 s}\,, \quad \sigma_{max} = \frac{Fh}{I}\frac{D}{2} = 4h\,\frac{F}{\pi D^2 s}$$

Hypothesis of the modification of shape energy:

$$\sigma_V = \sqrt{\sigma^2 + 3\tau^2}$$

$$\sigma_{allow} \ge \frac{F}{\pi D^2 s}\sqrt{16h^2 + 12a^2}$$

$$\Longrightarrow \quad s \ge \frac{2F}{\pi D^2 \sigma_{zul}}\sqrt{4h^2 + 3a^2}$$

Exercise: 59   **Chapter: 2.7,**   **Degree of Difficulty:** ✿✿*

A milling cutter, consisting of the shaft (radius r) and the cutter head (radius R), and clamped tightly in the chuck, is acted upon by force F during the cutting process at distance L from the tight clamping.

How large is the equivalent stress according to the hypothesis of the modification of shape energy at the point of the maximum load? (Hint: Shear stresses as a result of the lateral force are to be disregarded.)
Given: F, r, R=2r, L = 4r.

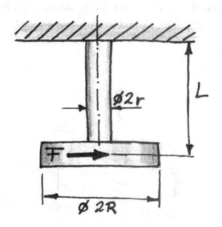

Solution

Overlapping: Torsion with $M_t = F R$
                          and   bending moment as a result of F.

Maximum load at the point of the maximum bending moment, i.e. at the place of tight clamping.

Bending stress:

$$\sigma = F L r / I = 4 \frac{FL}{\pi r^3} = 16 \frac{F}{\pi r^2}$$

Torsional stress:

$$\tau = M_t r / I_t = 2 \frac{FR}{\pi r^3} = 4 \frac{F}{\pi r^2}$$

Even state of stress:

$$\sigma_V = \sqrt{\sigma^2 + 3 \tau^2} = ... = \frac{4F}{\pi r^2} \sqrt{19}$$

A coil spring (radius R) is coiled from a wire (radius r, shear modulus G) with n coils lying close to one another. In an unburdened state it can be approximately assumed with R>>r that for the angle of inclination of the coils α ≈0. How large is the spring constant c of the coil spring?

Given: n, R, r, G.

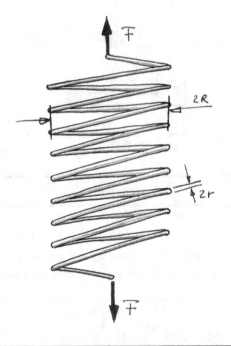

Solution

This is not so easy! And if we don't know at all what to do, we always start with a free-body diagram. And for the first coil it looks like this:

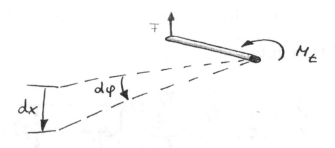

The cross-sections are burdened by a twisting moment $M_t = F\,R$. With this torsion, however, we can arrive at the classy torsion formula for the circular section:

$$d\varphi = \frac{2M_t}{G\pi r^4}\,ds\,,$$

where s is the coordinate that goes in the direction of the wire. In the free-body diagram, the resulting deformation $d\varphi$ is drawn with a hatched line. The displacement of the endpoint dx results from

$$dx = R\,d\varphi \quad .$$

The elongation ΔL of the total spring results then by way of

$$\Delta L = R \int_0^L \frac{2M_t}{G\pi r^4}\,ds = \frac{2FRR2\pi Rn}{G\pi r^4} \quad .$$

A comparison with the spring condition $F = c\,\Delta L$ yields a spring rigidity

$$c = \frac{Gr^4}{4R^3n} \quad .$$

Exercise: 61   **Chapter: 2.7,   Degree of Difficulty: 💣✳**

Two axles (diameter d_1, d_2, length L) are tightly clamped as depicted and are connected to one another free of backlash on the right side by two gear wheels (diameter D_1, D_2). How large is the contorsion of gear wheel 1 as a result of factor M?

Given: M, L, d_1, d_2, D_1, D_2, G.

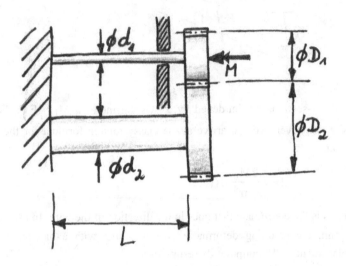

Solution

Central problem of the exercise: The cutting free of the gear wheels shows that, naturally, the same force F is acting upon the teeth of both gears. For axle 1 you then get:

$$M = M_{t1} + F\, D_1 / 2 \qquad \text{with } F = 2\, M_2 / D_2$$

$$= \frac{GI_{t1}}{L}\, \varphi_1 + M_2\, \frac{D_1}{D_2}$$

$$= \frac{GI_{t1}}{L}\, \varphi_1 + \frac{GI_{t2}}{L}\, \varphi_1 \left(\frac{D_1}{D_2}\right)^2$$

$$= \ldots \qquad = \varphi_1 \frac{G\pi}{32L}\left(d_1^4 + d_2^4\left(\frac{D_1}{D_2}\right)^2\right)$$

In order to adjust a bent sheet metal gutter (length L>>R, relationship of radiuses R/r=10, uniform thickness d, shear modulus G) the end cross-sections are momentarily rotated against one another by 90°. Which twisting moment is necessary for this and which maximum shear stress occurs here?

Given: L, r, R=10r, d, G, $\Delta\varphi$=90°.

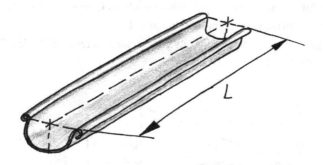

Cross section

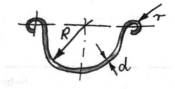

Solution

Torsional moment of inertia:

$$I_t = \frac{1}{3}\int_0^L d^3\,(s)\,ds \qquad \text{mit } L = 13\pi r$$

$$= 13\pi r d^3/3$$

Twisting moment:

$$M_t \quad = \quad \frac{GI_t \Delta\varphi}{L} = \frac{13\pi^2 rd^3}{6L}G$$

Shear stress:

$$\tau_{max} \quad = \quad M_t / W_t = \frac{G\pi d}{2L}$$

The cone tightly clamped on one side (shear modulus G, circular cross-section) is acted upon by twisting moment M_t at point x=L.

By what angle φ does the free end of the cone rotate as a result of the load?

Given: d, L<<d, G, M_t.

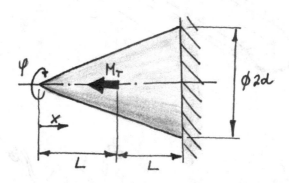

Solution

Rotation:

$$\Delta\varphi = \int\limits_{L}^{2L} \frac{M_t}{GI_t}\,dx = \frac{32M_t}{\pi G}\int\limits_{L}^{2L} \frac{dx}{D^4(x)} \quad \text{with } D(x) = d\,\frac{x}{L}$$

$$\Delta\varphi = \frac{32M_t L^4}{\pi Gd^4}\int\limits_{L}^{2L} \frac{dx}{x^4} = \dots = \frac{28}{3\pi}\frac{M_t L}{Gd^4}$$

The beam (round bar steel diameter d) tightly clamped at B is burdened by uniform line load q_0. How large is the vertical displacement of the beam's endpoint?

Given: q_0, d, L, E, G.

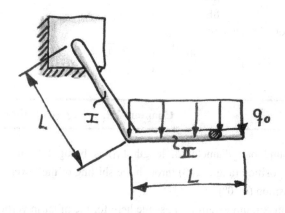

Geometrical moment of inertia:

$$I = \frac{\pi d^4}{64}$$

Torsional moment of inertia:

$$I_t = \frac{\pi d^4}{32}$$

Overlapping of three load cases:

1) Torsion beam I:

$$\varphi = \frac{M_t L}{G I_t}$$

$$= \frac{q_0 L^3}{2 G I_t}$$

293

==> displacement of the endpoint:

$$w_t = \frac{q_0 L^4}{2GI_t}$$

2) Bending of beam I: load case 1:

$$w_I = \frac{q_0 L^4}{3EI}$$

3) Bending of beam II: load case 3:

$$w_{II} = \frac{q_0 L^4}{8EI}$$

Fumbling around results in:

$$w_{TOT} = \frac{q_0 L^4}{\pi d^4}\left[\frac{16}{G} + \frac{88}{3E}\right]$$

Exercise: 65   **Chapter: 2.8,   Degree of Difficulty: ✿✲**

A pillar (round bar, diameter d, length L_0) is brought from the depicted tension-free position (angle $\alpha \neq 0$) through the shifting of the lower bearing into a vertical position ($\alpha = 0°$).

What is the maximum length L_0 possible here for the pillar in vertical position not to buckle?

Given: d, L=30d.

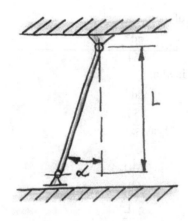

Buckle case 2: $F \quad = -EI\,\pi^2 / L^2$

Elongation: $\quad \varepsilon \quad = \quad (L-L_0)/L_0 = \dfrac{F}{EA} = -\dfrac{EI\pi^2}{L^2 EA}$

$\Longrightarrow \quad L - L_0 = \quad -L_0 \left[\dfrac{I\pi^2}{L^2 A} \right]$

$\Longrightarrow \quad L_0 \quad = \quad \dfrac{L}{1 - \dfrac{I\,\pi^2}{L^2 A}} \quad$ with $I = \dfrac{\pi d^4}{64}$ and $A = \dfrac{\pi d}{4}$

$\Longrightarrow \quad L_0 \quad = \quad 30.02\,d$

Exercise: 66 **Chapter: 2.8,** **Degree of Difficulty:** ✿✿

The push rod of a diesel motor's valve control is maximally acted upon by force F. The tube should be measured in such a way as to not transgress the maximum compression stress σ_{allow}, and there should additionally be a triple security against buckling.

How large do the inner diameter r and outer diameter R of the tube need to be? Given: F, E, L, σ_{allow}.

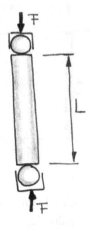

Compression stress:

$$\sigma_{\text{allow}} = \frac{F}{A} = \frac{F}{\pi(R^2 - r^2)}$$

Buckle case 2:

$$F = \frac{E\pi^2}{3L^2}\left[\frac{\pi}{4}(R^4 - r^4)\right]$$

I.e. two equations, two unknowns, solvable!

$$r = \sqrt{\frac{1}{2\pi}\left[\frac{12}{\pi}\frac{L^2\sigma_{\text{zul}}}{E} - \frac{F}{\sigma_{\text{zul}}}\right]}$$

$$\ldots \qquad R = \sqrt{\frac{1}{\pi}\frac{F}{\sigma_{\text{zul}}} + r^2}\ .$$

Exercise: 67   **Chapter: 2.8,   Degree of Difficulty:** 💣✳

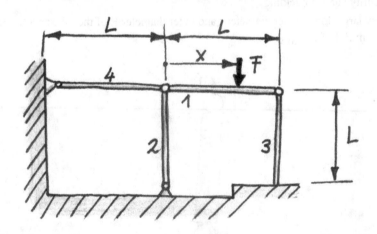

The depicted system consists of beam 1 (length L) and three equal, slim rods 2, 3 and 4 (each length L, bending rigidity EI).

At which point x of beam 1 does vertical force F have to act, if rods 2 and 3 are to have the same security against buckling?
(Hint: Deformations take place exclusively in the plane of projection!)
Given: L, EI.

Solution

Rod 2: Buckle case 2:

$$F_{crit2} = \frac{\pi^2 EI}{L^2}$$

Rod 3: Buckle case 3:

$$F_{crit3} = 2.0457 \frac{\pi^2 EI}{L^2}$$

ΣM around the point of load incidence, rod 1:

$$F_{crit2}\, x = F_{crit3}\, (L-x)$$

$$\implies \quad x = \frac{F_{krit3}}{F_{krit2} + F_{krit3}} L = \frac{2.0457}{1 + 2.0457} L \approx \frac{2}{3} L$$

Exercise: 68 **Chapter: 2.8,** **Degree of Difficulty: ✦✱**

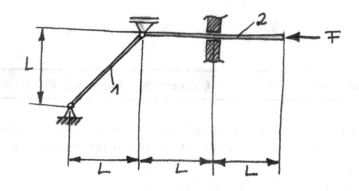

The depicted system is acted upon by force F.

How do the diameters d_1 and d_2 of the round bars have to be dimensioned in order for both bars to have the same security against buckling?
Given: L, E, F.

Solution

Rod 1: normal force:
$$F_1 = \sqrt{2}\ F$$

Buckle case 2:
$$F_{crit1} = \frac{EI_1\pi^2}{(\sqrt{2}L)^2}$$

Rod 2: normal force:
$$F_2 = F$$

right of the guide mechanism: buckle case 1,
left of the guide mechanism: buckle case 2
$\implies$ right critical point:
$$F_{crit2} = \frac{EI_2\pi^2}{4L^2}$$

$$\implies \quad \frac{F_{k1}}{F_1} = \frac{F_{k2}}{F_2} \quad ,$$

$$\frac{I_2}{I_1} = \sqrt{2}$$

$$\implies \quad \frac{d_2}{d_1} = \sqrt[8]{2}$$

Exercise: 69   **Chapter: 2.8,   Degree of Difficulty:** ✹

Two round bars with the same diameter are acted upon by force F as shown. The longer bar is supported in the middle by a thin rigid frictionless perforated metal plate.
a) How large is the buckle security of the system?
b) What buckle security does the system have if the perforated plate is removed?

Given: EI, a, F.

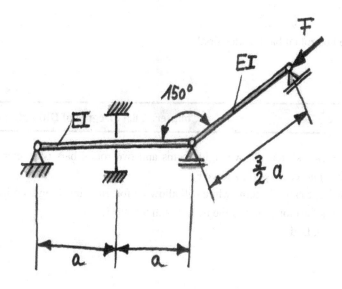

a)   The horizontal bar can be regarded as two bars of length a with buckle case 2. So we have three instances of buckle case 2 here. The slanted bar is more at risk here, since it is loaded with a larger normal force and is longer (3a/2) than either of the horizontal candidates for buckling

(a)   Buckle security:

$$S_k \quad = \quad \frac{F_{krit}}{F} = \frac{4\pi^2 EI}{9a^2 F}$$

b)   Horizontal bar:

$$S_k = \frac{F_k}{\frac{1}{2}\sqrt{3}F} = \frac{\sqrt{3}\pi^2 EI}{6a^2 F}$$

$$\frac{4}{9} > \frac{\sqrt{3}}{6},$$

i.e. the horizontal bar buckles first!

Exercise: 70   **Chapter: 2.8,   Degree of Difficulty: ◆※**

The depicted system of two rigid beams and two round bars (diameter d, E-module E) is acted upon by force F.

a) How large can F become while still allowing for triple buckling security?

b) What is the elongation of the tie rod with this load?

Given: a, d, E, d<<a.

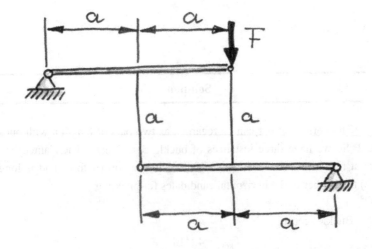

300

Statics: Cutting free of both bars: Normal force left is A, right is B.

ΣM upper beam:

$$Aa + B2a + F2a = 0,$$

ΣM lower beam:

$$A2a + Ba = 0,$$

$$\Longrightarrow \quad A = \frac{2}{3} F \text{ (tie rod)}, \qquad B = -\frac{4}{3} F \text{ (strut)}$$

Strut: buckle case 2:

$$-B = \frac{4}{3} F = \frac{1}{3} F_{crit} = \frac{1}{3} \frac{E\pi^2}{L^2} \frac{\pi d^4}{64}$$

$$\Longrightarrow \quad F_{max} = \frac{E\pi^3 d^4}{256 a^2} \quad .$$

Tie rod:

$$\Delta L \quad = \quad \frac{Na}{EA} \qquad \text{with} \quad N = \frac{2}{3} F_{max}$$

$$\text{and} \quad A = \frac{\pi d^2}{4}$$

$$\Longrightarrow \quad \Delta L \quad = \quad \frac{\pi^2 d^2}{96 a}$$

Exercise: 71   **Chapter: 2.8,   Degree of Difficulty: 💧✳**

The sketched connecting rod made from round material (length L, bending rigidity EI) is acted upon in operating direction by force F. Which distance a of bearing A should be chosen in order to achieve the highest possible resistance to buckling?

How large at most may F become in this case?

Given: L, EI.

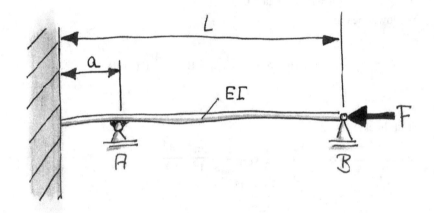

Solution

Now you've got that under your buckle... But how? First the load cases: In area I left of bearing A we have load case 4, in the right section II of the beam we've got load case 3. Now there's sure to be some protest. But we're sticking to our guns. And we also have an argument that will hopefully convince: The supposition often expressed by students and professors that we're dealing with cases 3 and 2 here would apply if the beam were interrupted by a joint at point A! For the continuous beam, a bending moment is transmitted at point A that must be included in the table for the chosen bearing form. Convinced?

The formulas from the table yield:

$$F_{\text{crit, I}} = 2.0457 \, \frac{EI\pi^2}{a^2} \,, \qquad F_{\text{crit, II}} = \frac{EI\pi^2}{(L-a)^2} \,.$$

The bearing is positioned best when the buckling load for both sections is equal:

$$2.0457 \, \frac{EI\pi^2}{a^2} = \frac{EI\pi^2}{(L-a)^2} \,.$$

Solving the equation that is quadratic in a yields after much scribbling

$$a = 0.59 \, L \,,$$

and now the load can be determined that will cause the beam to buckle:

$$F_{\text{crit, I}} = F_{\text{crit, II}} = 5{,}9 \, \frac{EI\pi^2}{L^2}$$

4.3. Kinetics-Kinematics

An old problem – perplexing theses [30]:

a) If force is always equal to counterforce, then no horse could ever pull a buggy, since the force of the horse, strictly speaking, would have to be canceled by the force on the horse. The buggy pulls the horse backwards with the same force as the horse pulls the buggy forwards.

b) But the horse definitely can pull the buggy forwards, since, as a result of losses, the horse pulls the buggy forwards with minimally more force than the buggy pulls the horse backwards.

c) Due to the limited reaction time of the buggy, the horse pulls the buggy forwards before the buggy can build up a counterforce.

d) The horse can only pull the buggy forwards if it is heavier than the buggy.

Which of the theses is/are correct?

None of the given theses are correct. Action is indeed equal to reaction - so the force that the horse exerts on the buggy is just as large as the force that the buggy exerts on the horse. But that's the only thing that's equal between the subsystems of horse and buggy. The force "wants to" cause a motion. The horse presses its hooves hard against the earth, since the force exerted by the buggy wants to pull the horse backwards. If its legs don't buckle or the horse doesn't slip (buggy too heavy), then the force would have to turn the entire earth backwards a bit. It will only have an infinitesimally small success in doing so. But the buggy is supported above the earth on wheels so that the force acting on the buggy only has to move the mass of the buggy!

Stated differently: If you observe the total system of buggy and horse, you will see that only the power exerted by the mud on the hooves (and the disregarded rolling friction of the wheels) acts on it in horizontal direction. And this causes, from a mechanical standpoint, the motion – of course, the horse has its part in the whole thing as well.

Exercise: 73   **Chapter: 3.1,   Degree of Difficulty: ✿**

Hinrichs, the scientist, lives completely isolated in his ivory tower that is represented here in simplified form as a box. The assumption is that the box moves at a constant velocity.

Dr. Romberg also isolates himself more and more – his box obviously rotates in a circle (thus he's attempting to balance out the rotation resulting from his dazed balance organ.)

Dr. Hinrichs   Dr. Romberg

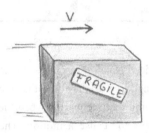   

Is it possible, with no contact with the outside world, to determine the state of motion of each box, for example the velocity [30], from inside the box?

Solution

Only the rotating gentleman can determine his state of motion. For example, if he lets his bottle fall, it will keep on moving tangentially to the circular path and thus won't land vertically under the drop point. The bottle lacks centripetal acceleration, or a force triggering a bending of the motion. If the acceleration is increased enough, the nausea detector in the stomach will make itself noticed at some point.

Dr. Hinrichs's box, however, is not accelerated. There are no forces at work here that are determined by the form of motion. Therefore, determining the form of motion is not possible. Even if you could look out of the box and see that the surroundings are moving in relation to the field of view, you couldn't say for sure whether you yourself are moving, whether your surroundings are being moved or whether both are shifted against one another. Anybody who's ever ridden on a train has experienced this. If the box were stopped or accelerated or shaken, you could feel the motion or test it with the impact test.

As a result, we become aware that:

A body either at rest or uniformly translationally moved is free of force, i.e. the acting forces cancel each other out.

Exercise: 74	Chapter: 3.2,	Degree of Difficulty: ☼

Two bike riders ride towards each other at a constant speed of 10 km/h. When they are exactly 20 km away from each other, a bee flies from the front tire of the right bicycle at an absolute velocity of 25 km/h directly to the front tire of the other bicycle. It quickly kisses the front tire (?), turns around in negligibly short time, and returns at the same velocity back to the first bicycle, kisses its front tire, turns around,... and keeps flying back and forth until both front tires crash into one another and the bee is squished.

What is the total distance (sum of the trips back and forth) that the bee has covered from the starting point on the front tire until the end of its life [30]?

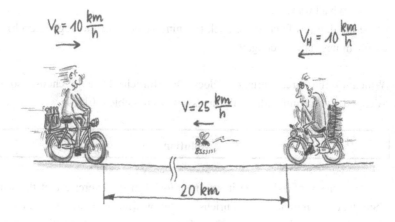

The solution for this exercise is really easy – if you follow the correct approach:

20 km at 2 x 10 km/h => Bicyclists' ride time: 1 h
Flight time of the bee: 1h
Velocity of the bee: 25 km/h => Flight distance: 25 km!

Exercise: 75   **Chapter: 3.2,   Degree of Difficulty:** ✿

Using pseudo-scientifically optimized methods, Dr. Hinrichs practices fetching with his dog. To do so, he goes walking for 15 minutes (walking speed 5 km/h) and throws a stick at a throw velocity of 10 km/h

   a) to the side,
   b) forwards,
   c) backwards.

Once the dog has fetched the stick (running velocity of the dog: 15km/h), the whole things starts over again.

Which of the three variations does Dr. Hinrichs have to choose so that Alexander von Humboldt[48] will run as long as possible [30]?

Solution

Oh, how nasty of us! Since it was mentioned in the exercise that the whole thing takes 15 minutes, Dr. Hinrichs can throw whichever way he wants – the dog still runs for 15 minutes. (The whole thing would've been different if we

---

[48] The name of Dr. Hinrichs's poor dog (Dr. Romberg has no comment)

had asked for which variation the dog runs the longest distance. But we didn't.) Here's a little tip for exams: Always think several times about what exactly is being asked! And when you're done, make sure that you've answered all parts of the question [30]!

Exercise: 76   **Chapter: 3,   Degree of Difficulty: ☼**

An Amtrak train speeds along at 400 cm/s despite a delay. Dr. Romberg, who "lost" his driver's license long ago, stumbles through the train in the direction of travel to the train's dining car (relative velocity to the train: 100 cm/s). While doing so, he stuffs without pause a hot dog into his mouth (feed rate of hot dog 5 cm/s). An ant walks along on the hot dog and, because of the alcohol fumes, flees from Dr. Romberg's mouth at a velocity of 2 cm/s. (see also [30]). What is the velocity of the ant relative to the track?

Variation:

Dr. Romberg relieves himself at v_{URINE} = 300 cm/s perpendicularly to the direction of travel through the hole in the toilet directly onto the track. How high is the velocity at which the stream collides with the ties (air resistance disregarded)?

Solution

First, to the intake of food:

The velocities of the train, the hot dog eater and the ant are rectified and can be added. The velocity of the hot dog is opposed to these velocities and thus has to be subtracted from the above-mentioned sum. So, for the absolute velocity you get

$$400 \text{ cm/s} + 100 \text{ cm/s} + 2 \text{ cm/s} - 5 \text{ cm/s} = 497 \text{ cm/s}$$

We observe: Rectified velocities can be added or subtracted.

Now on to the release of liquid:

In this case, the velocities aren't rectified. Here, the velocities have to be added vectorially.

So we look for the hypotenuse in a right triangle, where one leg corresponds to the train's velocity and the other leg corresponds to v_{URINE}.
For the absolute velocity, you end up with

$$v = \sqrt{400^2 + 300^2} \text{ cm/s} = 500 \text{ cm/s}$$

Exercise: 77	Chapter: 3.3, Degree of Difficulty: ✿

A boulder and a little stone of the same material fall the same height H with negligible air resistance. Which of the two objects flies for a longer time?

Solution

You can't say it often enough: Both require the same amount of time! This would be a good time to scribble out a free-body diagram. It's true that for the boulder, the accelerating force is much larger – but that's because a mass larger by the same factor has to be accelerated. Stated differently: If you crush the boulder into many little stones and let the whole heap fall, then nothing changes as far as the flying time is concerned. Thus, the resulting acceleration is equal in both cases.

Despite this fact, you hear again and again diverging commentary like, for example, while skiing: You got down to the bottom faster because you're heavier. This is generally not true. However, very precise investigations have shown that the friction relationships in the snow can be dependent on the normal force and really heavy skiers experience other friction relationships. But are these taken into consideration in the après-ski bragging?

Exercise: 78　　　　　　　**Chapter: 3.4,**　　　**Degree of Difficulty:** ☼

A stone is thrown into a muddy marsh. It penetrates 5 cm into the marsh. How fast does it have to be thrown in order to achieve a depth of penetration of 20 cm? (For all hardcore mechanics: The assumption is that the marsh consists of homogenous mud!) [30]

Solution

We only have to (surprisingly?) throw twice as fast. Common sense possibly leads us astray here. One could easily think: Four times the depth of penetration, i.e. four times the velocity. Humbug!

For four times the depth of penetration 4h, you need four times as much energy. You could also cut the marsh into 4 disks of the thickness h = 5 cm and requires the same amount of energy for each disk. The energy can be determined by

$$E = \int F ds = F h \quad ,$$

since in the case of the homogenous mud, the marsh's force of resistance remains constant. And where does the energy come from? From the kinetic energy $E = 0.5mv^2$ of the stone, of course. But this depends quadratically on the velocity. So: Four times the depth of penetration means double the velocity!

A rubber ball and a steel ball both have the same size, velocity and mass. Both are thrown against a shaky block [30].
a) Which of the balls is more likely to knock the block over?
b) Which ball causes more damage to the block?
(For all minimalists: A grounded explanation could also be very interesting!)

Solution

Before the ball's impact, the momentums of the various balls are equal. But the block's momentum is zero. After the impact of the ball, the steel ball and the block have a common (infinitely small) velocity. Here, the steel ball acts upon the block with a temporary blow of force.

In the case of the rubber ball, the momentum change is the same but up to two times as big as the momentum change of the steel ball, since the velocity doesn't just have to be slowed down, but also accelerated in the opposite direction. For this reason the rubber ball is much more likely to knock the block over. You could ask yourself: „But if I wanted to trash something, wouldn't I use a steel ball instead?" Yes, but for different reasons:

As far as the damage to be done to the block is concerned, let's examine our energy balance for a sec. The kinetic energy of the ball before the impact has to correspond to the kinetic energy of the ball after the impact as well as to the deformation energy of the block (and of the ball). In the simplest case, the energy at the same velocity turns back around from the block (ideal rubber ball). For the kinetic energy the sign of the velocity doesn't matter. For the energy balance, it means that no energy remains for the deformation of the block. With the steel ball, it's different: Here, the kinetic energy is used up, namely in the form of plastic deformations of the ball and of the block. Thus, the steel ball causes more damage in the form of plastic deformations!

Exercise: 80	Chapter: 3.3, Degree of Difficulty: ☼

Two people stand on a rotating disk (angular velocity ω). One person throws a ball at velocity v directly in the direction of the second person. At the moment of the throw, the second person is just passing a goal [30].

a) Does the ball hit the second person?

b) Does the ball hit the goal?

Solution

Unfortunately, we have to answer both questions with „No". If we determine the velocity of the ball, we find that it possesses velocity v in the direction of the center of the circle – and velocity ωR tangential to the disk as a result of the first person's rotation, where R signifies the distance of the thrower from the center of the circle. So, the ball flies further to the right than anticipated – and thus, to the right of the person as well as to the right of the goal. And to complicate the situation: In the time during which the ball – in the case of a correct throw directed further to the left that would arrive at the goal – the second person has already gone a bit further during the flight time. So hitting the second person is not that easy – that's why we'll also spare ourselves the calculation for the correct throw angle, ok?

If you observe the actual curve of the throw in the rotated subsystem's coordinate system – for example from the viewpoint of both people – then the ball apparently flies in a curved path. The observation of this deflection is all the more amazing if you don't notice that you are rotating (e.g. as inhabitants of locations other than the North and South Poles). The deflection is named after its discoverer Coriolis. If you wanted to devise a straight course in spite of this effect, you'd have to throw the ball into a tube extending between both people. The walls would bend the path's curve straight, exerting a force (coriolis force) onto the ball.

Exercise: 81 **Chapter: 3.1,** **Degree of Difficulty:** 💣※

Crank 1 turns at constant angular velocity ω. The rod rests on point P on the edge and does not lift away from the edge for the entire duration of the motion.

a) Determine the velocity pole for the depicted position.

b) Determine the angular velocity of rod 2 for φ=45°.

c) At which angles φ does rod 2 execute translational motions for a short time?

Given: ω, r.

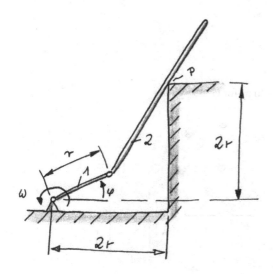

Solution

---

a) For the construction of the velocity pole, we know two velocity directions on object 2:

The velocity of the joint v_1 is vertical to the driving crank. Since rod 2 on the edge can neither lift off nor penetrate into the edge, v_2 is only possible in rod direction at point P. Following the rule according to Chap. 3.1, you get the velocity pole Q as the intersection of the perpendicular to v_1 and v_2.

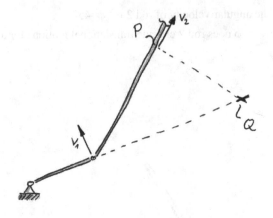

b) From the sketch analogous to a) for φ=45° yields

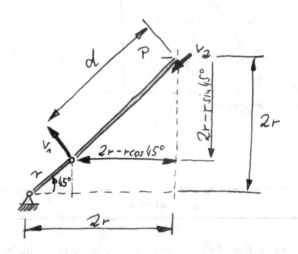

that the velocity pole lies in P. For the joint, which is a point on rod 2, the velocity v_1 is given by the crank via $v_1 = r \omega$.

But this point has – from the velocity pole of rod 2 – the distance

$$d = \sqrt{(2r - r\cos 45°)^2 + (2r - r\sin 45°)} = \ldots r(4 - \sqrt{2})/\sqrt{2}$$

So you get:

$$\omega_2 = v_1 / d = \omega \sqrt{2}/(4 - \sqrt{2})$$

c) A translational motion is given, when two velocities are aligned parallel to one another. In the given case it means that the crank must move in the direction of the rod. This is given for the depicted extreme positions I and II.

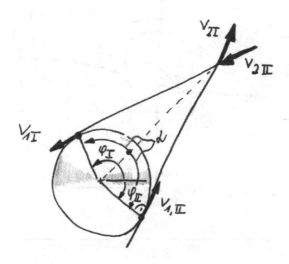

So we're dealing here with the situations for which the rod is tangentially oriented to the circle described by the crank endpoint. From the angular relationships, you get:

$$\cos\alpha = r/(2\sqrt{2}\,r) = 1/(2\sqrt{2}) \quad \Longrightarrow \quad \alpha = 69{,}3°$$

$$\varphi_{I,II} = 45° \pm \alpha,$$

$$\varphi_I = 114{,}3°, \quad \varphi_{II} = -24{,}3°$$

Exercise: 82   **Chapter: 3.1,   Degree of Difficulty: ✦✱**

The sketch show the drive unit of a steam engine that moves at velocity v_0. The wheels roll without slipping.

a) Determine for the sketched position the location of the connecting rod's velocity pole in the given coordinate system.

b) Determine the relative velocity of the piston relative to the cylinder.

Given: R, $r = \frac{1}{2}\sqrt{2}\,R$, v_0, L, $\alpha = 45°$.

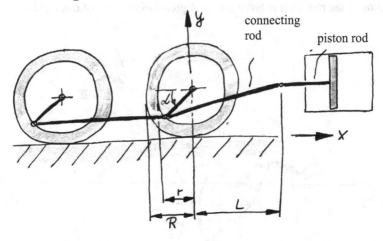

Solution

a) $x_Q = L,$ $y_Q = -L/\tan\alpha = -L$

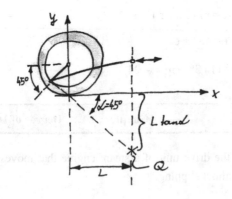

b) $v_{rel} = v_K - v_0$ with v_K: velocity of the piston rod

$v_K = \omega_S (R+L)$,

$\omega_S = v_S / d$ with $d = r + \sqrt{2}\, L = \dfrac{\sqrt{2}}{2} (R + 2L)$

$\omega_{wheel} = \dfrac{v_0}{R} = \dfrac{v_S}{r}$ $\Longrightarrow$ $v_S = \dfrac{r}{R}\, v_0 = \dfrac{\sqrt{2}}{2}\, v_0$

$\Longrightarrow$ $\omega_S \quad = \dfrac{v_0}{R + 2L}$

$\Longrightarrow$ $v_K \quad = \dfrac{R+L}{R + 2L}\, v_0$

$\Longrightarrow$ $v_{rel} \quad = \dfrac{-L}{R + 2L}\, v_0$

Exercise: 83   **Chapter: 3.2,   Degree of Difficulty:** ☼

A homogenous thin rod (length L, mass m) is first unstably balanced as shown and then falls over as a result of a slight disturbance.

What is the velocity v_{end} of the free rod end at the point in time at which the rod passes the stable balanced state (rod hangs downwards)?

Given: L, g.

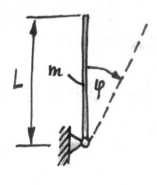

319

energy theorem (zero level: bearing):

$$mgL = 0.5\, J\, \omega^2 \qquad \text{with} \quad J \quad = \quad \tfrac{1}{3} m\, L^2$$

$$\text{and} \quad v_{end} \quad = \quad \omega\, L$$

$$\Longrightarrow \quad v_{end} = \sqrt{6gL}$$

Exercise: 84   **Chapter: 3.2,**   **Degree of Difficulty:** ✿

From a tower of height H, within the earth's gravitational field a ball is thrown vertically upwards at initial velocity v_0. Determine

a) the maximum throw height H_{max} and the rise time t_{max},
b) the time t_E until the ball hits the earth again.
Given: $H = 5$ m, $v_0 = 10$ m/s, $g = 9{,}81$ m/s².

Solution

a) $\dot{y}(t_{max}) = 0 = -g t_{max} + v_0 \quad \Longrightarrow \quad t_{max} = v_0 / g = 1{,}02\,s$

         $y(t_{max}) \quad = -\dfrac{1}{2} g t_{max}^2 + v_0 t_{max} = 5{,}1$ m,

         $H_{max} = y(t_{max}) + H = 10{,}1$ m

b) $y(t) \quad = -\dfrac{1}{2} g t_E^2 + v_0 t_E = -H$

      $\Longrightarrow \quad t_E^2 - \dfrac{2v_0}{g} t_E - \dfrac{2H}{g} = 0$

      $\Longrightarrow \quad t_E = 2{,}45$ s.

A cylinder (mass m, radius R) lies on a horizontal plane. The friction coefficient between plane and cylinder is $\mu = \mu_0 = 0.1$. The cylinder is set in motion by means of the depicted force $F = 0.1\,mg$.

a) Does the cylinder roll or slip?

b) What is the acceleration of the cylinder?

Given: m, R, F=0.1mg, $\mu = \mu_0 = 0.1$.

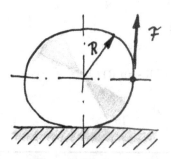

Solution

Free-body diagram:

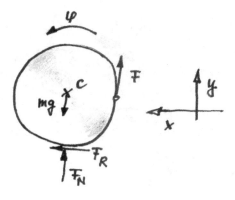

Determination of F_N:

$$\Sigma F_y = m\,\ddot{y} = 0 = F + F_N - mg, \qquad ==> \quad F_N = 0.9\,mg$$

$$\Sigma F_x = m\,\ddot{x} = F_R$$

$$\Sigma M^C: \quad J^C\,\ddot{\varphi} = F\,R - F_R\,R \qquad \text{with } J^C = 0.5\,m\,R^2$$

Assumption: The cylinder rolls! (assumption valid if $F_R < \mu\,F_N$)

$$==> \quad \ddot{x} = R\,\ddot{\varphi}$$

Connecting the equations together:

$$\ldots \quad F_R = \tfrac{2}{3}\,F \quad < \mu\,F_N$$

==> Assumption valid !

Determination of angular acceleration:

$$\ddot{\varphi} \quad = (F\,R - F_R\,R)\,/\,J^C = \frac{2g}{30R}$$

$$\ddot{y} \quad = \ddot{\varphi}\,R = \frac{1}{15}\,g$$

Exercise: 86   **Chapter: 3.2,**   **Degree of Difficulty:** ✿

After enjoying some local delicacies, Dr. Romberg leaves the mountain cabin.

On the way back, the doctor, in high spirits, reaches the depicted position in his car (mass m, velocity v_0), turns his complete attention to the stars, shifts into neutral and rolls... after distance L into a wall (spring constant c). What it the maximum force exerted on the wall?
Given: c, m, g, L, α, v_0.

Solution

Choice of the zero level[49]: maximum downward deflection of spring of the wall.

Energy theorem (x is the downward deflection of spring of the elastic wall):

$$m\,g\,(L+x)\sin\alpha + 0.5\,m\,v_0^2 = 0.5\,c\,x^2$$

Engineer-like simplification: $x \ll L$

$$\Longrightarrow \quad x = \sqrt{\frac{2mgL\sin\alpha + mv_0^2}{c}}$$

$$F = c\,x = \sqrt{c(2mgL\sin\alpha + mv_0^2)} \quad .$$

Exercise: 87   **Chapter: 3.3,   Degree of Difficulty: ◓✳**

A ball (mass m, radius R) lies on a block (mass 2m) that lies frictionlessly on a plane. The block is accelerated by force F – simultaneously, the ball begins to roll. What is the friction force between the ball and block?

Given: m, R, F.

---

[49] (Note: Dr. Romberg's level: ZERO is independent of the location)

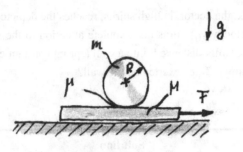

First, the free-body diagrams:

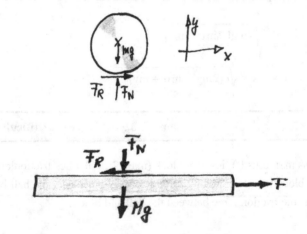

Dr. Romberg starts practicing first with the egg-holder on a book: „This is hilarious – first it turns backwards and then in the other direction! Funny, huh?"

Theorem of linear momentum for the subsystems:

$$\Sigma F_{x, \text{ball}} = m \, \ddot{x}_{C,\text{ball}} = F_R$$

$$\Sigma F_{x, \text{block}} = M \, \ddot{x}_{\text{block}} \qquad = F - F_R$$

Theorem of twist for ball:

$$\Sigma M_{\text{ball}}{}^C = J_{\text{ball}}{}^C \, \ddot{\varphi} = F_R \, R \text{ with } J_{\text{ball}}{}^C = \tfrac{2}{5} mR^2$$

Kinematics:

$$\ddot{x}_{\text{block}} = \ddot{x}_{C, \text{ball}} + \ddot{\varphi} \, R$$

And now combine everything (4 equations, four unknowns):

$$\text{....} \quad F_R = F / 8$$

Exercise: 88   **Chapter: 3.3,**   **Degree of Difficulty:** ♦※

A ball (mass m, radius R) that turns at angular velocity ω_0 is set on a plane (friction coefficient μ).
a) How long does it take before pure rolling begins?
b) Which horizontal velocity has the ball reached by that point?
Given: m, R, ω_0, μ.

Solution

Theorem of linear momentum for the ball that has been cut free:

$$\Sigma F = m \, \ddot{x} \text{ center of gravity}_1 = F_R = \mu \, mg$$

Theorem of twist for the ball around the center of gravity:

$$\Sigma M = J \, \ddot{\varphi} = -F_R \, R \text{ with } J = \tfrac{2}{5} m R^2$$

$$\implies \quad \ddot{\varphi} = \dot{\omega} = -\frac{F_R R}{J} = \text{const}$$

$$\implies \quad \omega(t) = \omega_0 + \dot{\omega} \, t = \omega_0 - \frac{F_R R}{J} t$$

Condition for the transition into rolling state:

$$\dot{x} = \mu\, gt = \omega(t)\, R \quad \ldots \quad \Longrightarrow \quad t = \frac{2\omega_0 R}{7\mu g}$$

$$\Longrightarrow \quad \dot{x} = \tfrac{2}{7}\,\omega_0\, R$$

(Dr. Hinrichs can also calculate the energy that's lost in rolling, i.e. not converted into motion: It's supposedly 5/7 of the initial energy)

A pendulum (length L) suspended from a massless string and possessing mass m is let go from the depicted horizontal position. The string breaks at the point in time at which the string force is $F_F = F_{crit} = 1.5\, mg$. What velocity does the mass possess at this point in time?

Given: m, $F_{crit} = 1.5\, mg$, L, g.

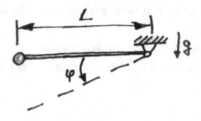

Solution

Free-body diagram:

Theorem of linear momentum in string direction:

326

$$\Sigma F \quad = \quad F_F - mg\sin\varphi = m\,\omega^2\,L$$

Energy theorem:

$$0.5\,(mL^2)\,\omega^2 \quad = \quad mgL\sin\varphi$$

Combine with one another:

$$\varphi = 30°, \qquad v = \sqrt{gL}$$

A cylinder (mass m, radius r) rolls (almost) with no initial velocity down the depicted hill. At angle $\varphi = \varphi_0$ the cylinder begins to slip.
What is the static friction coefficient μ?
Given: m, r, R, φ_0.

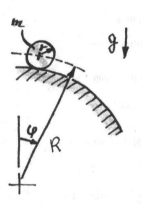

Solution

Free-body diagram:

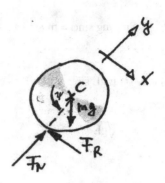

Theorem of twist around the center of gravity:

$$0.5 \, m \, r^2 \, \ddot{\psi} = F_R \, r$$

Theorem of linear momentum:

$$\Sigma \, F_x : m \, r \, \ddot{\psi} = mg \, \sin\varphi - F_R$$

Combine:

$$R = \tfrac{1}{3} \, mg \, \sin\varphi$$

Theorem of linear momentum :

$$\Sigma \, F_y : F_N - mg \, \cos\varphi = - \, m \, \frac{v^2}{R}$$

Energy theorem with roll condition:

$$g \, R \, (1 - \cos\varphi) = \tfrac{1}{2} \, m \, v^2 + \tfrac{1}{2} \, \tfrac{1}{2} \, m \, r^2 \, \frac{v^2}{r^2} = \tfrac{3}{4} \, m \, v^2$$

Combine:

$$F_N = \tfrac{1}{3} \, mg \, (7 \cos\varphi - 4)$$

Law of friction at the point in time of the transition from rolling to slipping:

$$F_R = \mu \, F_N$$

$$\Longrightarrow \quad \sin\varphi_0 = \mu \, (7 \cos\varphi_0 - 4)$$

$$\Longrightarrow \quad \mu = \frac{\sin\varphi_0}{7 \cos\varphi_0 - 4}$$

Dr. Romberg has once again enjoyed himself immensely during an evening in his favorite bar. He is subsequently carried home in a trailer attached to Dr. Hinrich's car. Determine the force in the rod connecting car and trailer coupling during a downhill stretch with the brakes engaged, assuming that all of the wheels are still rolling without slipping. (mass car + Dr. Hinrichs: m_1, mass trailer, Dr. Romberg + beer bottles: m_2)

Given: $m_1, m_2, g, \alpha, \mu_1, \mu_2$.

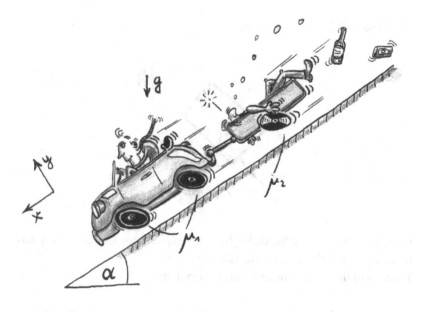

Solution

Dr. Hinrichs's free-body diagram:

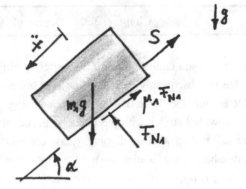

Dr. Romberg's free-body diagram:

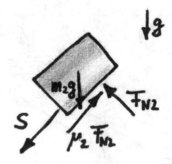

First, we will adhere to the daring basic assumption that Dr. Romberg's mass remains constant during the entire ride.

Theorem of linear momentum for the total system:

$$\Sigma F_X = (m_1 + m_2) g \sin\alpha - m_1 g \cos\alpha \, \mu_1 - m_2 \, g \cos\alpha \, \mu_2$$
$$= (m_1 + m_2) \, \ddot{x}$$

$$\Longrightarrow \quad \ddot{x} = g \sin\alpha - \frac{m_1}{m_1 + m_2} g \cos\alpha \, \mu_1 - \frac{m_2}{m_1 + m_2} g \cos\alpha \, \mu_2$$

Theorem of linear momentum , e.g. for mass 1:

$$\Sigma F_X = -S + m_1 \, g \sin\alpha - m_1 g \cos\alpha \, \mu_1 = m_1 \, \ddot{x}$$

Application of $\ddot{x}$:

$$\Longrightarrow \quad S = \frac{m_1 m_2}{m_1 + m_2} \, (\mu_2 - \mu_1) \, g \cos\alpha \qquad .$$

A Ferris wheel (radius R) rotates at the constant angular velocity ω. Dr. Hinrichs's female love interest (mass m) sits in the lower gondola, and Dr. Romberg's old flame (mass M = 2m) is in the upper gondola. For inexplicable reasons, both are sitting on a scale. How fast (sought: ω) does the Ferris wheel have to turn in order for both scales to show the same weight in the depicted position?[50]

Given: g, R.

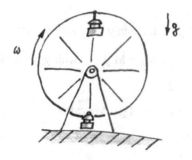

Solution

Here's a little note for your confusion: The gondolas move ONLY translationally, even though they're moving in a circle... Go ahead and rack your brain over that!!! (Little tip: Sketch the velocity directions for two points of a gondola as opposed to a radial rod.)

The free-body diagram looks like this:

---

[50] This exercise seems to be taken from real life, doesn't it?

Sum of the forces:

$$\Sigma F_y = F_{N1} - mg = m\,\ddot{y} \qquad \text{with } \ddot{y} = \omega^2 R$$

resp. $\quad \Sigma F_y = F_{N2} - Mg = M\,\ddot{y} \qquad \text{with } \ddot{y} = -\omega^2 R$

Condition: $\quad F_{N1} = F_{N2}$

$\Rightarrow \quad m\,\omega^2 R + m\,g = M\,g - M\,\omega^2 R$

Solution: $\quad \omega = \sqrt{\dfrac{g}{3R}}$

Exercise: 93   **Chapter: 3.3,   Degree of Difficulty: ♦**

Dr. Hinrichs, riding a bicycle (common center of gravity C, total mass m, velocity v_0), has to brake suddenly (friction coefficient of street + tires: μ), since the drunk Dr. Romberg has sat down in the middle of the street.

How long is the brake path if Dr. Hinrichs only uses the back wheel brake and the back wheel is not yet at the point of locking?

Given: m, g, v_0, μ.

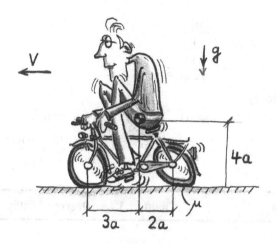

Free-body diagram:

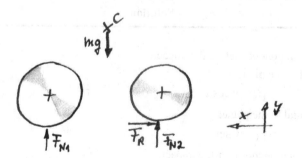

Theorem of linear momentum :

$$F_x = m\,\ddot{x} = -\mu\,F_{N2}$$

$$F_y = m\,\ddot{y} = 0 = F_{N1} + F_{N2} - mg$$

Theorem of twist around the center of gravity:

$$0 = -3a\,F_{N1} + 2a\,F_{N2} + 4\,a\,\mu\;F_{N2}$$

Combine:

$$\Longrightarrow \quad F_{N2} \quad = \frac{3mg}{5+4\mu}$$

$$\Longrightarrow \quad \ddot{x} \quad = -\frac{3\mu}{5+4\mu}\,g$$

Course of velocity:

$$v(t) \quad = v_0 + \ddot{x}\,t \quad \Longrightarrow \quad t_{stop} \quad = -v_0\,/\,a$$

$$s(t_{stop}) = v\,t_{stop} + \frac{\ddot{x}}{2}\,t_{stop}{}^2 = -\frac{v_0{}^2}{2a} = \frac{(5+4\mu)v_0{}^2}{6\mu g}$$

Exercise: 94   **Chapter: 3.4,**   **Degree of Difficulty:** ✿◐❋

A ball falls vertically onto an inclined plane (number of collisions e). What does the angle of inclination of the plane have to be in order for the ball to fly away horizontally after the impact?

Given: e.

Solution

Velocity components before the impact:
1) Normal to the plane:

$$v_{1N} = v\cos\alpha$$

2) Tangential to the plane:

$$v_{1T} = v\sin\alpha$$

Velocity components after the impact:
1) Normal to the plane:

$$v_{2N} = -e\,v\cos\alpha$$

2) Tangential to the plane:

$$v_{2T} = v\sin\alpha$$

Determination of the component in vertical direction:

$$v_{2V} = v_{2N}\cos\alpha + v_{2T}\sin\alpha$$

Condition for ball to fly away horizontally:

$$v_{2V} = 0 = -e\,v\cos\alpha\,\cos\alpha + v\sin\alpha\,\sin\alpha$$

Reformulate with

$$\cos^2\alpha = 1 - \sin^2\alpha$$

$$an^2\alpha = \frac{\sin^2\alpha}{1 - \sin^2\alpha}$$

$$\Rightarrow \quad \tan\alpha = \sqrt{e}$$

Exercise: 95   **Chapter: 3.4,   Degree of Difficulty:** ✦✶

The fist of a karate fighter, mass m = 0.05 kg, chops a board of mass M = 5 kg in two. The impact velocity v_0 is 600 m/s, the fist leaves the board at velocity v = 150 m/s. The board can be shifted frictionlessly on the base.

a)   What is the velocity V of the board after the chop?

b)   How much energy is lost in the damaging of the board?

Given: m = 0.05 kg, M = 5 kg, v_0 = 600 m/s, v = 150 m/s.

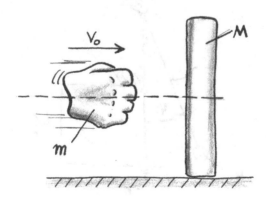

Conservation of momentum must apply:

$$m \, v_0 = m \, v + M \, V$$

$$\Rightarrow \quad V = m \, (v_0 - v) / M = 4.5 \text{ m/s}$$

Energy balance:

$$.5 \, m \, v_0^2 = 0.5 \, m \, v^2 + 0.5 \, M \, V^2 + E_{broken}$$

$$\Longrightarrow \quad E_{broken} = 0.5 \, m \, v_0^2 - (0.5 \, m \, v^2 + 0.5 \, M \, V^2) = 8386{,}9 \text{ kgm}^2/\text{s}^2$$

Exercise: 96   **Chapter: 3.4,   Degree of Difficulty:** ✿⬥※

A pendulum consists of a massless string and a mass $M = 1$ kg attached to the string at a distance of $L = 2$ m. The pendulum is released from the initial deflection $\varphi_0 = 30\,°$ and collides at $\varphi = 0$ with another pendulum (mass $m = 0.5$ kg, $e = 1$) with a length of only $L/2$.

How far does the shorter pendulum swing following the impact?

Given: $M = 1$ kg, $m = 0.5$ kg, $L = 2$ m, $\varphi_0 = 30°$, $e = 1$, $g = 10$ m/s².

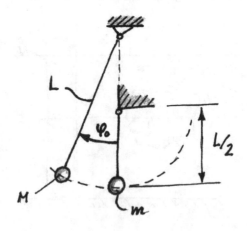

Velocity before the impact:

$$g L (1-\cos\varphi_0) = 0.5 J \omega^2 \qquad \text{with } J = M L^2$$

$$\Rightarrow \quad \omega = \sqrt{\frac{2g(1-\cos\varphi_0)}{L}} , \qquad v = \omega L = ... = 2.3149 \text{ m/s}$$

Velocity of the second pendulum after the impact:

$$V = \frac{1}{m + M} [(1+e) M v] = ... = 3.0866 \text{ m/s},$$

$$\Longrightarrow \quad \omega_2 = 2V/L$$

Energy theorem after the impact:

$$0.5 J_2 \omega_2^2 = m g L (1-\cos\varphi_{max}) / 2 \qquad \text{with } J_2 = m L^2 / 4$$

Solve for φ_{max}:

$$\varphi_{max} = \arccos\left(1 - \frac{J_2\omega_2^3}{mgL}\right) .$$

Combining of the equations, apply:

$$\varphi_{max} = 58{,}4°$$

Exercise: 97   **Chapter: 3.4,**   **Degree of Difficulty:** ◆※

A bouncing ball strikes at points P_1, P_2, P_3, P_4 .. against a (ideally smooth) plane. The distance between the impact points P_1 and P_2 is the same as the distance between points P_2 and P_4. What is the number of collisions e? Given: $|P_1 P_2| = |P_2 P_4|$.

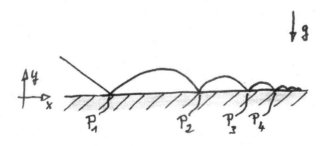

The x-component of the velocity remains constant during the impacts ($v_{x0} =$ const); in y-direction, the equations for the free fall can be used:

$$y(t) = v_{y0}\, t - 0.5\, g\, t^2$$

Time between two impacts:

$$y(T) = 0 \qquad \Longrightarrow \qquad T = \frac{2 v_{y0}}{g}$$

Distance d between two points of impact:

$$d = 2\,\frac{v_{x0} v_{y0}}{g}$$

Neighboring distances:

$$d_{n+1} = e\, d_n$$

$$d_{n+2} = e^2\, d_n$$

$$|\,P_1\, P_2\,| \qquad = |\,P_2\, P_4\,|$$

$$d_n \qquad\quad = d_{n+1} + d_{n+2}$$

$$1 \qquad\qquad = e \quad + e^2$$

$$\Longrightarrow \quad e = 0.618$$

References

[1]   B. Assmann, Technische Mechanik, Band 1: Statik, München, Wien: R. Oldenbourg Verlag, 1984.

[2]   B. Assmann, Technische Mechanik, Band 2: Festigkeitslehre, München, Wien: R. Oldenbourg Verlag, 1979.

[3]   B. Assmann, Technische Mechanik, Band 3: Kinematik und Kinetik, München, Wien: R. Oldenbourg Verlag, 1985.

[4]   D. Besdo, B. Dirr, B. Heimann, K. Popp, I. Teipel, Formelsammlung zu Technische Mechanik I-IV, Institute of Mechanics, University of Hannover.

[5]   D. Besdo, B. Dirr, B. Heimann, K. Popp, I. Teipel, Collection of Examination Exercises I-II, Institute of Mechanics, University of Hannover.

[6]   D. Besdo, B. Dirr, B. Heimann, K. Popp, I. Teipel, Collection of Examination Exercises III-IV, Institute of Mechanics, University of Hannover

[7]   Beitz, W.; Küttner, K.-H. (Herausgeber): Dubbel, Taschenbuch des Maschinenbaus, 15. Auflage, Berlin, Heidelberg, New York, Tokyo: Springer Verlag, 1986.

[8]   Euler, L., Theorie der Bewegung fester oder starrer Körper (Theoria motus corporum solidorum seu rigidorum), Greifswald: C. A. Koch's Verlagsgesellschaft, 1853.

[9]   H. Göldner, F. Holzweißig, Leitfaden der Technischen Mechanik: Statik, Festigkeitslehre, Kinematik, Dynamik, Darmstadt: Steinkopff Verlag, 1984.

[10]   H. Göldner, D. Witt, Lehr- und Übungsbuch Technische Mechanik, Band 1: Statik und Festigkeitslehre, Fachbuchverlag Leibzig Köln, 1993

[11]   D. Gross, W. Hauger, W. Schnell, Technische Mechanik, Band 1: Statik, Berlin, Heidelberg: Springer Verlag, 1988.

[12]   W. Hauger, W. Schnell, D. Gross, Technische Mechanik, Band 3: Kinetik, Berlin, Heidelberg: Springer Verlag, 1986.

[13]   G. Holzmann, H. Meyer, G. Schumpich, Technische Mechanik, Teil 1: Statik, Stuttgart: Teubner Verlag, 1990.

[14]   G. Holzmann, H. Meyer, G. Schumpich, Technische Mechanik, Teil 2: Kinematik und Kinetik, Stuttgart: Teubner Verlag, 1986.

[15] G. Holzmann, H. Meyer, G. Schumpich, Technische Mechanik, Teil 3: Festigkeitslehre, Stuttgart: Teubner Verlag, 1983.

[16] Istituto Geographico de Agostini S. p. A., da Vinvi, L., Das Lebensbild eines Genies, Wiesbaden, Berlin: Emil Vollmer Verlag.

[17] K.-D. Klee, Elastostatik, Lecture Manuscript, University of Applied Science Hannover, 1. Auflage, 1994.

[18] K. Magnus, H. H. Müller, Grundlagen der Mechanik, Stuttgart: Teubner-Verlag, 1990.

[19] E. Mönch, Basic Lecture about Mechanics, Munich, Vienna: R. Oldenbourg Verlag, 1981

[20] Newton, I., Mathematische Prinzipien der Naturlehre, mit Bemerkungen und Erläuterungen von Prof. Dr. J. Ph. Wolfers, Berlin, Verlag von Robert Oppenheim, 1872.

[21] Pestel, E., Technische Mechanik, Band 1: Statik, Mannheim, Wien, Zürich: BIVerlag, 1982.

[22] E. Pestel, J. Wittenburg, Technische Mechanik, Band 2: Festigkeitslehre, Mannheim, Wien, Zürich: BI Verlag, 1983.

[23] E. Pestel, Technische Mechanik, Band 3: Kinematik und Kinetik, Mannheim, Wien, Zürich: BI Verlag, 1988.

[24] Ritter, A., Theorie und Berchnung eiserner Dach- und Brücken-Konstruktionen, Hannover: Carl Rümpler, 1863.

[25] Ritter, A., Lectures about Mechanics, 1860.

[26] W. Schnell, D. Gross, W. Hauger, Technische Mechanik, Band 2: Elastostatik, Berlin, Heidelberg: Springer Verlag, 1989.

[27] I. Szabó, Repetitorium und Übungsbuch der Technischen Mechanik, Berlin, Göttingen, Heidelberg: Springer Verlag, 1963.

[28] I. Szabó, Einführung in die Technische Mechanik, Berlin, Heidelberg, New York: Springer Verlag, 1975.

[29] I. Szabó, Geschichte der mechanischen Prinzipien, Basel, Boston, Stuttgart: Birkhäuser Verlag, 1987.

[30] Epstein, Lewis Caroll: Thinking Physics. Practical Lessions in Critical Thinking, Insight Press, San Francisco

[31] D. Labuhn, O. Romberg: Keine Panik vor Thermodynamik!, Verlag Vieweg, 2005

[32] J. Strybny: Ohne Panik Strömungsmechanik!, Verlag Vieweg, 2003

[33] Telephone Book of Maui, Hawaii (incl. classified directory), Edition 2006

340

Index